Heating with COAL

By John W. Bartok, Jr.

GARDEN WAY PUBLISHING
Charlotte, Vermont 05445

Allen County Public Library
Ft. Wayne, Indiana

Copyright 1980 by Garden Way, Inc.

All rights reserved. No part of this book may be reproduced without permission in writing from the publisher, except by a reviewer who may quote brief passages or reproduce illustrations in a review with appropriate credit; nor may any part of this book be reproduced, stored in a retrieval system, or transmitted in any form or by any means—electronic, photocopying, recording, or other—without permission in writing from the publisher.

Cover photo by George Robinson

Illustrations by Robert Vogel

Printed in the United States

First Printing, November 1980
Second Printing, October 1981

Library of Congress Cataloging in Publication Data

Bartok, John W   1936–
  Heating with coal.

  Bibliography: p.
  Includes index.
  1. Heating. 2. Coal. 3. Stoves, Coal. I. Title.
TH7400.B37   697′.042   80–24941
ISBN 0–88266–243–0 (pbk.)

CONTENTS

Introduction 1

Why Choose Coal? 2
Should Energy Conservation Be First? 7
Will a Coal Stove Change Your Life-Style? 10
Environmental Impact 11
What About Safety? 13

CHAPTER 1
Coal as a Fuel 15

Formation 15
Peat 16
Lignite 17
Subbituminous 17
Bituminous 18
Cannel Coal 18
Anthracite 19
Coke 19
Mining 20
Physical Properties 24
Buying Your Coal 27
Kindling Needed 31
Where to Store It 31
Selecting the Proper Coal 33
How Coal Burns 34

CHAPTER 2
Selecting a Stove 37

Location 38
Size 39
Radiant Stoves 41
Circulating Stoves 42
Coal in a Wood Stove 43
Materials for Stoves 44
Important Features 46
Appearance 51
Stove Efficiency 52
Heating Part of Your Home 54
Heating Most of Your Home 59
Used Stoves 61
The Stove Shop 62
Accessories 63
False Advertising 64
Checklist for Selecting a Stove 65

CHAPTER 3
67 A Safe Installation
- 68 Who Will Install the Stove?
- 68 Hiring a Professional Installer
- 69 Stove Permit—Is It Needed?
- 70 Hearth or Stoveboard
- 72 Protecting the Wall
- 76 Stovepipe
- 79 Chimneys
- 84 Using the Fireplace Flue
- 89 A New Chimney
- 95 Heat Distribution
- 95 Make-up Air
- 97 Mobile Home Stoves
- 99 Checklist for a Safe Coal Stove Installation

CHAPTER 4
101 Firing the Stove
- 102 What Size and Type of Coal?
- 103 Operating the Stove
- 104 Burning Anthracite
- 109 Burning Coke
- 110 Burning Bituminous
- 113 Burning Subbituminous and Lignite
- 114 Burning Coal in a Fireplace
- 116 Ash Disposal
- 116 Creosote
- 117 Stove Maintenance
- 118 Your Stove During the Summer
- 119 Problem Stoves

CHAPTER 5
123 Central Heat
- 123 Equipment Definitions
- 124 How to Decide
- 125 Furnace Design
- 128 Furnace Installation
- 130 Boiler Design
- 132 Boiler Installation
- 134 Multi fuel Furnaces and Boilers
- 135 Automatic Coal Burners
- 138 Draft Requirements
- 139 Hand Firing
- 139 Maintenance
- 140 Converting a Converted Furnace Back to Coal

143 Concluding Thoughts
145 References
147 Glossary
153 Appendix
157 Acknowledgments
159 Catalog of Manufacturers
185 Index

INTRODUCTION

As we hold a lump of coal in our hand and look at the imprints of ferns and twigs we wonder what the earth looked like over 300 million years ago as these materials began their long transformation. The dense, very lush vegetation that grew in the large swamps was good cover for the amphibians and reptiles that inhabited the earth at that time. The accumulation of this matter over the years and the movement of the earth's surface led to the formation of coal. The energy that the plants absorbed then is now millions of years later a heat source for our homes and industry. This form of energy is one of our greatest resources today. As the reserves of oil, another more recent organic material, decline and the supplies of easily obtained wood decrease, coal is starting to return as an important home heating fuel.

Before discussing coal let's look briefly at several reasons for installing alternate energy systems to heat our homes. The homeowner who faces ever-mounting fuel bills may be forced to consider one of the alternate fuels. The homes of the pre-1930 era with many large rooms and little or no insulation require large amounts of fuel. Closing off rooms as the children leave for school or get married will help some but the central heating system is usually not set up to provide control of individual areas. A switch to a coal or wood stove in three or four rooms that are "lived in" makes good sense.

Even in our more modern homes the energy costs can be a significant part of the budget. Until recently, home builders did not consider the installation of adequate insulation a requirement for selling the home. The extra money was usually put into the interior finish or an energy-using fireplace. For those building or buying a new home, making provisions for the stove installation before rather than after the home is built is very important. Eliminating the fireplace and putting the money into the stove is a wise choice.

Feel Secure

Security may be another reason. If you have five or six tons of coal in your bin as you enter the heating season you know that whatever the world energy situation is during the winter you and your family will stay warm. Many areas of the country are subject to snow and ice storms with subsequent power outages lasting several days. A coal stove can help you through this emergency by supplying your heating and cooking needs.

Some people like a challenge. The challenge of learning a method of heating that was practiced by our forebearers is exciting. The daily ritual developed is different for each stove installation. Most stove users quickly become "experts" and often are found discussing "their method" with their neighbors.

The American people have a strong sense of patriotism. We quickly join together to overcome external forces that may threaten our way of life. As the outflow of money continues to the oil-producing countries many people are installing stoves to utilize our own country's natural resources.

Why Choose Coal?

In this book we will be discussing coal as an alternate fuel. There may be references to wood from time to time as the stove selection process and the installation are basically the same. Also, many homeowners are considering a switch from wood to coal and the differences in firing these fuels need to be pointed out.

Wood makes a good fuel if it is well seasoned and is a sound hardwood. If you have access to a productive woodlot and own the proper equipment, procuring the wood may be easy. But many homeowners get their wood from someone else's property. After a year or two the easy-to-get wood along the roadways is depleted and you end up carrying it a hundred yards or more to the truck or trailer.

Hard Work

Getting wood also involves a lot of hard work. Each cord weighs between two and three tons and is usually handled at least six times before it is placed in the stove. For some people this may be a form of relaxation, for others it's a way to keep in shape, but for most it's drudgery.

This nation has large reserves of coal. The 1974 report of the U.S. Department of Interior, Bureau of Mines indicates that the total identified

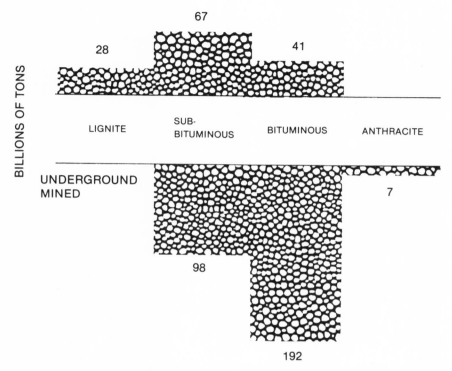

This chart shows the coal reserves of this country that are available using present technology.

coal reserves in the United States amount to 1600 billion tons, of which 434 billion tons are considered economic to mine at today's prices and with present technology. The 1979 total production was only 0.6 billion tons. Even if consumption increases several times the present output, the reserves should last several hundred years. Although the reserves of anthracite, the most common domestic fuel, are less than the more common bituminous, Bethlehem Mines, one of the larger companies, has reserves of 500 million tons and is mining only a half million tons annually.

Plentiful supplies of coal are found near most of the population centers of the country. Supplies from the mines have been adequate to keep up with the increasing demand.

Getting your winter's supply involves calling your local coal dealer. Some dealers have been in business many years and date back to the coal era of the early 1900s. Many new suppliers are starting in business. These see the increasing demand for coal and are trying to fill this need. Many of these are in the urban-city areas where the price of wood is exorbitantly high and coal is a much less expensive alternate fuel. Most dealers carry several sizes of coal.

Harder to Burn

Coal is harder to burn than wood, due to its higher ignition temperature, about 200°–400° higher. It also requires different techniques for firing and some wood burners have a difficult time making this change. On the other hand, a good coal fire in a properly designed stove requires less attention, only a shaking and refueling once or twice a day. Also in contrast to wood, coal generally does not develop much creosote that can cause a dangerous chimney fire.

The amount of coal needed for the heating season will vary with the size of your home, its location, how well it is insulated, and several other factors. Table 1 will give an approximation of the tons of coal needed to replace your present heating fuel. Remember to subtract any fuel used for water heating, cooking, or other accessory need before you multiply.

Your winter's supply of coal can be stored in a bin in your basement or under cover outside. It requires about one-third the space of wood of an equivalent heat value. A space about twice the size of a 275-gallon oil tank will hold five tons, enough to heat an average, well-insulated home all winter.

TABLE 1. APPROXIMATE COAL EQUIVALENT OF PRESENT FUEL*

Multiply gallons No. 2 fuel oil by 0.0075 to get —— tons coal
Multiply cubic feet natural gas by 0.00006 to get —— tons coal
Multiply kilowatt hours electricity by 0.00025 to get —— tons coal
Multiply cords air dried hardwood by 0.83 to get —— tons coal

* Efficiencies used: Refer to fuel cost comparison chart, page 5.

Example: If your last winter's fuel consumption (excluding that used for heating domestic hot water) was 1000 gallons of No. 2 fuel oil, you would require 1000 × 0.0075 = 7.5 tons of coal.

Comparing Cost to Other Fuels

The cost of coal as compared with other fuels is an important consideration. A check with local dealers can get you the current price. Coal is sold in bulk in ton quantities and bagged in 10-, 25-, and 50-pound

FUEL COST COMPARISON CHART

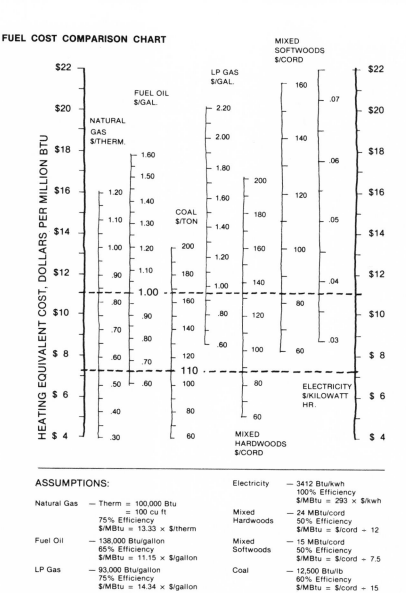

ASSUMPTIONS:

Natural Gas	— Therm = 100,000 Btu = 100 cu ft 75% Efficiency $/MBtu = 13.33 × $/therm	Electricity	— 3412 Btu/kwh 100% Efficiency $/MBtu = 293 × $/kwh
Fuel Oil	— 138,000 Btu/gallon 65% Efficiency $/MBtu = 11.15 × $/gallon	Mixed Hardwoods	— 24 MBtu/cord 50% Efficiency $/MBtu = $/cord ÷ 12
LP Gas	— 93,000 Btu/gallon 75% Efficiency $/MBtu = 14.34 × $/gallon	Mixed Softwoods	— 15 MBtu/cord 50% Efficiency $/MBtu = $/cord ÷ 7.5
		Coal	— 12,500 Btu/lb 60% Efficiency $/MBtu = $/cord ÷ 15

One Btu = The amount of energy required to raise a pound of water 1° Fahrenheit. 8.3 Btu will raise one gallon 1° F.

To use the chart, read across the fuel price columns to the Heating Equivalent column to determine the price per MBtu (Million Btu).

For example, if a source of coal is available at $110./ton, the equivalent fuel cost is approximately $7.25/MBtu. Fuel oil at $1.00/gallon has an equivalent fuel cost of $11.10/MBtu. In this case, it may pay to switch to coal because coal costs less per MBtu than oil.

Adapted from "Comparing Heating Fuel Costs" Bulletin FS-12, Northeast Regional Agricultural Engineering Service, Cornell University, Ithaca, NY 14853.

sacks. It is generally $30 to $40 a ton cheaper in bulk. The accompanying chart allows you to compare your present fuel and some of the alternate fuels on a heat equivalent basis. In some cases you may find your present fuel is less expensive than coal. A comparison of the heat values of different fuels is shown in the illustration. The therm is a common fuel quantity used throughout the industry. In this example the useful heat based on average efficiencies is used.

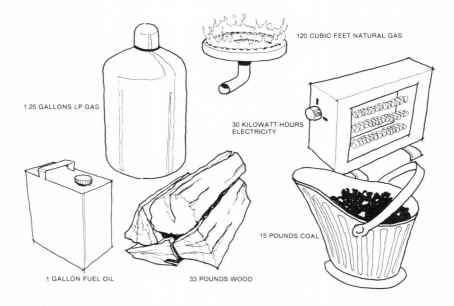

One therm equals 100,000 Btu. It also represents the heat derived from these sources.

What is the installation going to cost? Several factors will affect that price. Location is one. The stove should be placed where it will distribute heat through the home. Because the stove is a point source of heat, those rooms nearer the stove will be warmer. This can be used to advantage when you want to have your living areas warmer than the bedrooms. The location will affect such costs as for stovepipes and wall and floor protection. And of course the size (and the cost) of stoves vary. Further discussion on selection and installation will be covered in later chapters.

Another major cost factor is the chimney; where it is located or where it can be located. A new chimney will cost from $300 to $2000. If you have an extra flue in your chimney or can use a fireplace flue, you may be able to eliminate this large expense. It's important that the flue lining be

TABLE 2. COST OF INSTALLING YOUR COAL STOVE

Item	Range of Cost	Your Cost
Stove	$300–$900	_____
Floor protection	25–75	_____
Chimney (if needed)		
Masonry	400–2000	_____
Factory-built	$1–$1.50/in. of length	_____
Wall protection		_____
Other (stovepipe, stove accessories)		_____
	Total Cost	_____

sound and in good repair. You can develop your own cost figures for a stove installation using the guide in Table 2.

Should Energy Conservation Be First?

If you are interested in installing an alternate energy system you should also be interested in energy conservation. As mentioned before, when the price of fuel was low many of our older homes were built with a minimum amount of insulation. Unless you have modernized your home or have recently built an energy-efficient home, the best use of your money is in energy conservation measures. These items have a relatively short payback period (generally one to four years) and some of the cost can be deducted as Internal Revenue Service tax credits. By weatherizing your home before installing the coal stove, your present fuel usage will decline, and you will use less coal when you install your stove.

Here are some goals to make your home more energy efficient.

Insulation. As heated air tends to rise, much of the heat we have in our homes leaves through the attic. The map gives the inches of ceiling insulation recommended for your home. The type of insulation and how it is installed is important. A good discussion of this is given in the booklets "In the Bank or Up the Chimney? A Dollars and Cents Guide to Energy Saving Home Improvements" and "Home Energy Savers' Workbook." Both are available from the U.S. Government Printing Office, Washington, D.C. 20402, or from most state energy offices. A list of these offices can be found in the Appendix. Another good book is *547 Easy Ways to Save*

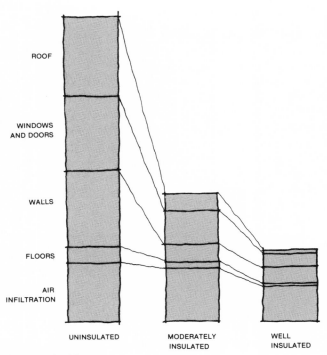

Heat losses from typical houses

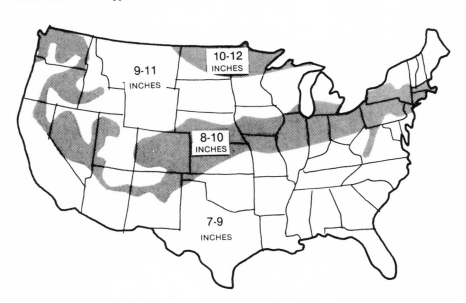

Depth of ceiling insulation you should have in various parts of the country to avoid costly heat losses through the attic.

Energy in Your Home (Garden Way Publishing). These are also good sources of information on alternate energy.

It is difficult to insulate the walls of an existing home. Door and window headers, fire stops, and wall plugs create barriers for the free flow of blown or poured insulation. In most cases a vapor barrier also is needed on the winter warm side of the insulation. A competent insulating contractor can determine whether this is feasible.

Basement insulation should be considered if you plan to put your stove there. Heat loss through these uninsulated walls can cost you an extra ton or more of coal each winter. Rigid foam material covered with gypsum board or paneling is generally the best material to use.

Storm windows. The unprotected window areas of your home lose as much as ten times that of the insulated walls. Double or triple glazing or

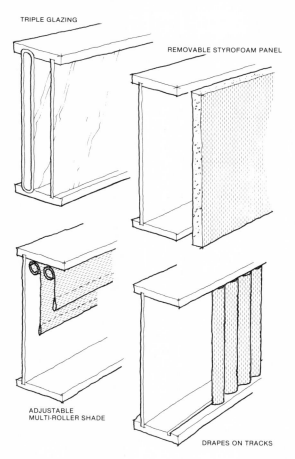

Four methods of window insulation

thermal curtains can reduce this considerably. Making your own storm windows out of clear vinyl plastic can help reduce the cost.

Infiltration. Cracks around doors, windows, and foundations allow the cold outside air to leak in during the winter and the warm inside air to escape. You notice this more on windy days when it feels colder in the house. Fuel is needed to heat the air that enters. Caulking, weatherstripping, and sealing the area at the sill of the house can pay a high return if done properly. The materials used are inexpensive and the work can be done by the homeowner with a few hand tools.

Heating system. A gas or oil heating system should be cleaned, adjusted, and checked for efficiency at least once a year. The furnace or boiler in most homes runs for 1000 hours or more each heating season. With the increasing cost of fuel even a small increase in efficiency can give you a significant savings. Other ways to save fuel include replacement of obsolete oil-fired units with new heat-retention burners that can reduce fuel consumption 10 to 20 percent; the installation of an automatic flue damper that stops the flow of heated air escaping when the furnace is off and can save an average of 10 percent; and insulation for ducts and pipes in unheated areas to reduce heat loss.

A key consideration in purchasing a new or used home should be its energy efficiency. Before buying, inspect it carefully. It's well to remember that once built, a home cannot be economically brought up to maximum energy conservation standards although significant savings can be achieved.

Will a Coal Stove Change Your Life-Style?

It takes a special family to tolerate some of the inconvenience of a coal stove. If you become serious about using coal you must commit at least half an hour a day to the operation and maintenance of the stove. A daily routine may consist of rising a little earlier to fire up the stove so that the house is warm when the rest of the family gets up. During the day an occasional check is probably all that is needed. At night the fire needs to be banked and a safety check made to see that no combustibles have been left near the stove. Also once a day you must remove the ashes, a rather messy job.

More Dirt, Dust

You can also expect to have more dirt and dust in the house. Coal tends to chip and break, forming dust particles. Many of these are very fine and become airborne, attaching themselves to drapes, rugs, and furniture, especially if the stove is located in the living area. More frequent vacuuming may be necessary to keep the house clean.

On the plus side—if you have a young growing family a coal stove in the living or family room can help to bring you closer together. Most families find that the kids do their homework and play near the warmth of the stove. The many hours spent near the stove can help us understand our children better.

Environmental Impact

If everyone starts burning coal what will happen to our air quality? This question is commonly asked by those concerned with the increasing use of wood and coal as alternate fuels.

In *Coal—Bridge to the Future* (Ballinger Publishing Company, Cambridge, MA, 1980) it was concluded that although the levels of some pollutants may increase with greater use of coal, additional research is needed to determine the long-term effects. Coal must be compared with other energy alternatives that have their own environmental, health, and safety effects.

THE TWO-FUEL SYSTEM

There are two seasons of the year—spring and fall—when it's difficult not to supply more than the required amount of heat with a coal stove.

During the day the temperature rises to about 60° F., but it drops to below freezing at night. It's a time when heat is needed only part of the day. But building a coal fire, only to let it go out each day, is a waste of your energy.

That's when many homeowners switch to small wood fires, in the evening and perhaps again in the morning. The wood fire is simple to start and to keep going as long as heat is needed.

This two-fuel system works well with most coal stoves but, as we'll explain, simply doesn't work with a wood stove not built for burning coal.

Sulfuric Acid

The volatile matter in coal consists mainly of hydrogen, nitrogen, oxygen, and sulfur. The first three of these are found in the air we breathe. Sulfur in coal occurs as iron sulfide and the organic sulfur and some sulfates. Both the iron sulfide and the organic sulfur burn to the gases sulfur dioxide, the rotten egg gas, and sulfur trioxide, a reactive colorless gas. About one-third of the sulfur burns to sulfur trioxide. At temperatures below 300° F., the temperature often found in chimneys, this can combine with the water vapor driven off the coal to form sulfuric acid. This acid, mainly from power plants and nitrogen oxide emissions from autos, is mixed with the air and the rain carries it into the water supply. Scientists have established that this is killing fish in some Eastern lakes.

How much does domestic heating add to this pollution problem? A typical 800-megawatt electricity generating plant operating at an average 75 percent capacity consumes 2.6 million tons of coal a year. It would take more than 500,000 homes in the northern climates consuming an average five tons each to equal this consumption. It is true that power plants have scrubbers on the chimneys limiting the amount of pollutants exhausted, but still the pollution from 2.6 million tons of soft coal is significant. The coal commonly used in home heating has a sulfur content lower than the emission level from the power plant.

Coal differs from seam to seam and from region to region. The amount of impurities can vary from as little as 2 percent in some anthracite to as

---

WOOD OR COAL?

Advantages of Wood	Advantages of Coal
Cheaper in rural and suburban areas close to supplies.	Cheaper than wood in most cities, and in many rural areas.
Easier to start fire.	Coal takes less space to store.
Quicker heat.	Relatively free of creosote.
Cleaner to burn.	Easier to keep fire going for long periods.
Ashes valuable to garden.	
Coal stoves must be built better, so are more costly than wood stoves.	Doesn't have to be cut, split, dried or seasoned.
Wood supply is aesthetically pleasing.	

much as 20 percent in some forms of lignite. This mineral matter, which can contain numerous chemicals, turns to ash when the coal is burned. Some of this leaves the chimney or settles in the stovepipe as fly ash, but most of it falls through the grate and is collected in the ash pan.

Because of the possible presence of some undesirable heavy metals, coal ashes should not be spread on the garden.

What About Safety?

You can learn to enjoy using your coal stove and save some on your heating bill, but if you burn the house down all is for naught. Safety with coal stoves comes under three areas: first, an approved stove; second, a proper installation; and third, safe operation of the stove. Today most manufacturers have submitted their stoves to a testing agency for review. Each stove is subject to several firing rates including an extreme test called a *flash fire test*. Most stoves pass these tests with only slight modifications to their designs.

Once you purchase the stove, you have the responsibility of seeing that it is installed according to either the National Fire Protection Association or the Underwriters Laboratories code. You can have the store install it for you or you can install it yourself. The latter is not difficult if you are handy with tools. You will have to be careful about installing it a safe distance from combustible materials including furniture, being sure that there is adequate clearance around the stovepipe where it enters the chimney, and making certain the chimney is tight and clean. It's also important to have someone check your work to see that you haven't forgotten something. The local building inspector or fire marshal is usually the best choice. In some states a building permit is required. Some insurance companies also require that they be notified so that they can inspect the installation. We will cover the subject of a safe installation in greater detail in Chapter 3.

You can see that burning coal involves a commitment from your family and considerable thought and planning. For those who have decided to install a stove, the remaining chapters in this book will go into greater detail on selecting, installing, and using your stove.

CHAPTER 1
COAL AS A FUEL

Coal is a combustible rock which had its origin in the accumulation and partial decomposition of vegetation. It is estimated that the oldest of the coals is over 300 million years old and formed during the carboniferous era. These are found mainly in North America, Asia, and Europe. Other types of coal date to about 100 million years ago.

Formation

The most popular theory on how coal was formed is that the earth was covered with a dense growth of lush vegetation consisting of trees, vines, and fern-like plants. The climate during this period was probably tropical with high humidity and warm temperatures. There were ample nutrients in the soil and the plants grew rapidly. Another contributing factor may have been a high carbon dioxide (CO_2) level. During this period there may have been many large volcanic eruptions and the atmosphere may have had an excess of CO_2, greater than the 300 parts per million (ppm) that we have today. Extra carbon dioxide benefits some types of plants and commercial greenhouses use it as a supplement to increase growth. For example, lettuce production can be increased about 50 percent by adding CO_2 to a level of 1500 ppm. The extra CO_2 during this period may have allowed a greater amount of sunlight to reach the earth with a resulting decrease in the radiation of heat from the soil.

The climate of the Coal Age was probably fairly uniform with few cold areas anywhere in the world. Coal has even been located in the Antarctic.

With the ideal climate for vegetative growth it is likely that each year a thick vegetative growth mat was formed. It has been estimated that twenty feet of vegetative matter are required to form one foot of coal. Such a buildup probably took three centuries.

Pressed by Glaciers

Pressure, needed to compress all this vegetative matter, probably came from the huge glaciers that covered parts of the earth. Volcanic eruptions and earthquakes may have also occurred, causing shifts in the earth's surface. These shifts spread water and sediment across large sections of the earth. The impurities usually found imbedded in coal probably were introduced then.

Coal usually is found in sloping beds separated by layers of clay. These *seams* are anywhere from a few inches to as much as several hundred feet thick. Some of the beds cover hundreds of square miles.

Evidence supporting this theory of formation are the many imprints of leaves and twigs found in coal. Fossilized trees have also been found upright in coal beds.

Peat

The first step in the transformation process to coal is the formation of *peat*. This is the decaying vegetative matter and is usually brown and very fibrous. Peat can be found throughout the world in bogs and swamps. It is commonly used as fertilizer or soil amendment for growing crops.

Peat is used as a fuel in some parts of the world where wood and coal are in short supply. Because of its very fibrous nature it holds a lot of moisture and must be dried before it can be burned.

Lignite

The next step is the formation of *lignite*. This is the first material to be ranked as a coal in most of the classification systems (Table 3).

Lignite is formed by the compression and burial of peat with a sedimentary material. Areas in the lower Mississippi Valley, Texas, Montana, and the Dakotas contain large lignite deposits. Its color varies from brown to a deep black. Its heat value is reduced somewhat by its high moisture content. When it is mined it may be as much as 45 percent moisture. Drying can reduce this to around 20 percent. This is similar to some types of wood such as ash. Lignite is also similar to wood in that it has a high volatile matter content, is relatively easy to ignite, and burns with a great deal of smoke. When used as a fuel it should be fired with a thick bed and strong draft.

TABLE 3. SIMPLE RANKING OF COAL

Lignite
Subbituminous
Bituminous
Subanthracite
Anthracite

Subbituminous

This coal is found in most of the same areas as bituminous coal. Extensive deposits are located in Alaska and the Rocky Mountain east slopes.

The lower ranks of subbituminous contain up to 50 percent natural bed moisture. This is one of the deterrents to its greater use. One characteristic of this coal is its tendency to ignite spontaneously when stored in piles over ten feet high. This is caused by rapid physical and chemical changes that take place when the temperature fluctuates.

Subbituminous coal is used for domestic heating in some areas of the West. It has a medium heat value and can be fired so there is very little smoke.

Bituminous

The largest reserves of coal in the United States are of the bituminous class often called *soft coal*. The areas with the greatest reserves are in the Appalachian region, Mississippi Valley, and Colorado.

Large quantities are used for power generation, steel production, and the manufacture of coke.

Bituminous coals can be grouped into several classifications from a high-volatile moist coal to a low-volatile dry coal. The higher the volatile content generally the greater the heat value.

Bituminous coals are termed soft coals because they feel silky to the touch and can be broken easily into various sizes. The classes with the highest volatility tend also to be dusty. Some soft coals have a glassy luster, others are very dull.

Cannel Coal

This variety of bituminous coal is a specialty coal and probably had a different origin. Formed from deposits of stagnant water containing gelatinous algae, fish, and other animal remains, cannel coal contains

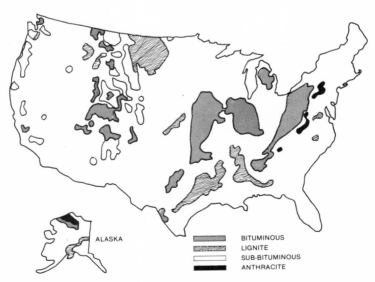

Here are where the nation's coal reserves are found. Map is adapted from *Coal Facts*, National Coal Association, Washington, D.C. 20036.

much volatile matter. It derives its name from the Welsh word for candle. Up to 20 percent of its weight is in the gas. During the 1800s it was used for its illuminating effect in fireplaces.

Cannel coal should not be used in a stove because it expands upon heating and often causes puffbacks from the spontaneous combustion of the trapped gas. Most of this country's cannel coal is located in Kentucky, West Virginia, and Utah.

Anthracite

For many years after it was discovered in the United States, anthracite was considered a rock and not a fuel. In 1766 it was first used as a heat source for a blacksmith's forge. Today it is considered a premium fuel for home heating, foundries, and electrical components.

Anthracite is the oldest and hardest of the coals. It is high in carbon content and low in volatile matter. Although it is more difficult to ignite, it burns longer than the other types of coal. Because of its low volatile content it is smokeless.

Pennsylvania contains most of the anthracite reserves in the United States. Much smaller deposits are found in Virginia, Arkansas, Colorado, and Washington. A 1000-square-mile deposit of anthracite and meta-anthracite (less than 2 percent volatile matter) is located in the Narragansett field of Rhode Island and Massachusetts. This coal is extremely variable in its character, has a high ash content, and is difficult to ignite.

Coke

Coke is often used as a home heating fuel. It is bituminous coal that has been heated in an oven or retort to over 1000° F. for ten hours or more and had the volatiles driven off. These by-products are captured and used in a variety of ways. *Coal gas*, or *coke-oven gas* as it is sometimes called, is used as a heating fuel in many cities. It is often mixed with other gases and piped to the home. The heat output is 400–600 Btu per cubic foot. Other products captured include coal tar, benzene, ammonium sulfate, and sodium cyanide.

The coal used for making coke must be of the coking type to form a porous compact mass. Systems used in production include the beehive oven and the more modern closed oven. Coke may be burned with little or no smoke and is important in metallurgical work for heating. The heat value is about the same as bituminous coal on a per-pound basis but only 1200–1400 pounds of coke can be obtained from a ton of coal.

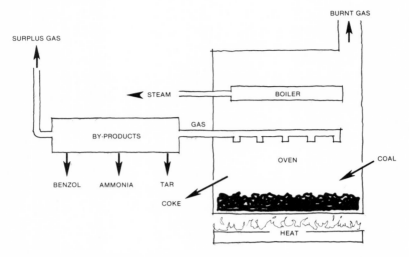

In making coke, bituminous coal is heated in an oven at temperatures of more than 1000° F. for ten or more hours. The volatiles are driven off, captured, and converted into tar, ammonia, benzol, and surplus gas. A ton of coal converts to 1200 to 1400 pounds of coke.

Mining

This country has about 4600 mines in operation employing from one to more than 1000 miners. The fifteen largest mining companies produce over one-half of the coal mined.

Many factors must be considered before an area is selected for development as a mine. These include the quality of the coal, size, depth, and location of the seam, and its location relative to transportation. In the eastern United States the coal is roughly 300 million years old, and the seams have been compressed to an average of four to six feet. The coal, mostly bituminous, is of a high heat value (10,000–12,000 Btu/lb.) and medium sulfur content (2–3 percent). Although some of the coal is near the surface, most of it lies several hundred feet below the surface.

In the West, the coal is about 100 million years old, and the seams average ten to thirty feet thick. This low sulfur coal (less than 1 percent) is of a lower heat content, generally 7,000–9,000 Btu/lb. Most of the production is by surface mining as the reserves are near the surface.

Underground (deep) mining. Mining is a complex process involving very expensive equipment and methods developed over many years. About 50 per-

Coal as a Fuel **21**

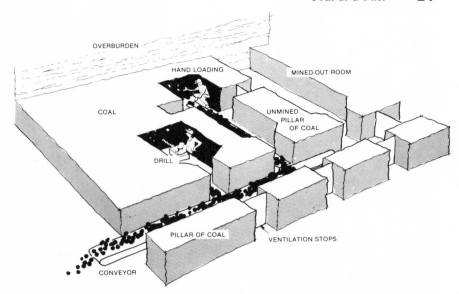

In underground mining, miners carve out "rooms" divided by pillars of coal that support the rock layer above. Coal in the pillar area is removed as the mine is being closed.

cent of the coal comes from deep mines. Sometimes the coal is found 1000 feet below the surface.

A deep mine resembles a series of rooms with the rock layer above the coal seam supported by pillars as the coal is removed. The coal in the pillar area may be mined as the mine is being closed.

Surface (strip) mining. As deep mining is considered one of the most hazardous occupations, greater emphasis has been placed on developing surface mines. In areas where the overburden (material over the coal seam) is relatively shallow, the trees, soil, and rocks are removed to expose the seam of coal.

The large shovels and draglines that have been developed during the past few years have helped to make surface mining economical. The bucket on these machines can scoop up to 200 yards of soil in one bite. That's a volume equal to one-half of a 24-by-36-foot house. These machines cost from three to seven million dollars each.

Strip mining tends to produce more of the less desirable fine coal. It also causes greater environmental damage from the water that leaches some of the undesirable chemicals from the spoil area and the erosion of banks before they have been replanted.

Huge shovels like this one are used in strip mining.

A coal breaker in Pennsylvania.

Coal as a Fuel   **23**

Sizing and cleaning. The coal is removed from the mine, transported to the processing plant, then graded for size, and the rocks and other minerals removed. The sizing consists of running the coal through a crusher and then over a series of screens. Screen openings vary in size with the type and location of the coal but generally fit into the grades shown in Table 4. The impurities can be removed by several methods. In some plants, larger pieces are removed by hand as they are carried by conveyor

TABLE 4. SIZES OF ANTHRACITE COAL

	Test Mesh Size			Test Mesh Size	
	Over	Through		Over	Through
Broken	3 in.	4 3/8 in.	Pea	9/16 in.	13/16 in.
Egg	2 5/16	3	Buckwheat	5/16	9/16
Stove	1 5/8	2 5/16	Rice	3/16	5/16
Nut	13/16	1 5/8	Barley	3/32	3/16

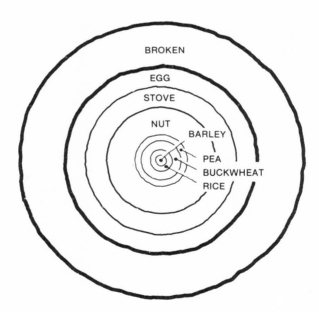

SIZES OF BITUMINOUS COAL

The screening and sizing practices vary widely. The following are typical of those used in stoves and furnaces.

	Test Mesh Size	
	Over	Through
Egg	2 in.	5 in.
Nut	1 1/4	2
Stoker	3/4	1 1/4

past a sorting station. Newer methods utilize the difference in density between coal and the impurities. Removal of the impurities helps to reduce ash content and transportation costs.

Iron pyrite, the main sulfur-bearing material, can also be removed through gravity separation. Pyrite has a density of 300–325 lbs./ft.³ compared to 75–84 lbs/ft.³ for bituminous coal. The sulfur content can be reduced 20 to 50 percent in most coals.

Some coal is washed to remove the dirt and dust. It may then be dried and coated with calcium chloride, petroleum oil, wax, or other material to make it less dusty.

Physical Properties

Several other characteristics of coal should be mentioned as they affect the way different coals burn.

Ash content. Mineral matter such as dirt, stone, and clay was mixed in when coal was formed. Traces of sulfates, carbonates, and phosphates are also present. The color of the coal ash usually indicates the impurity found in the greatest amount. A whitish ash signifies clay and a reddish ash means larger quantities of iron.

The ash in coal is valueless mineral matter that you pay for at coal prices. Good coal will have an ash content in the 2 to 5 percent range after processing. Most coals have a much higher content when mined. Wood, on the other hand, has from 0.5 to 2 percent. This means that you get two

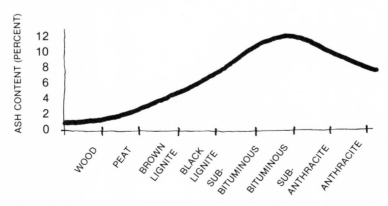

The average ash content of various fuels. The values of coal given are averages as mined. After processing, the ash content for the coals is usually from 2 to 5 percent.

to three times as much ash from coal as from an equivalent amount of wood, and you will have to dump the ash pan more often, usually daily.

Excess ash content also affects the operation of the stove. You will have to shake the grate more often to reduce the ash level. Greater deposits will be found in the stovepipe and chimney cleanout. The larger the size of coal used, generally, the less the amount of ash in the coal.

Friability. The tendency of the coal to break or crumble when being handled is called *friability*. Anthracites are generally the least friable, with the soft types of bituminous being the most.

This characteristic is important to the stove operator in that a less friable coal means less dust and dirt. It also means that you will have fewer fines (very small pieces) when the coal dealer dumps a two-ton load into your basement.

Moisture content. Water in coal must be considered both for its weight and its effect in stove performance. Unlike wood, which has a high moisture content even when air dried, the harder types of coal have very little inherent moisture. Both anthracite and bituminous as delivered generally contain less than 2 percent moisture. Even when stored outside they absorb very little and remain essentially dry.

Lignite and subbituminous, on the other hand, average about 25 percent as mined and must be dried before being burned. These should be stored under cover.

The average moisture content of various fuels. The values given are for air-dried fuel.

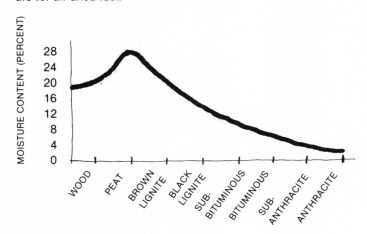

Heat value. This is the most important property of coal. It is made up of the fixed carbon and the volatile matter. As can be seen in the graph of properties of various solid fuels, the hard coals are almost pure carbon with very few volatiles. This gives a long, even burn cycle. Soft coals tend to be lower in carbon and higher in volatiles. This makes for easier starting and quicker pick up when the damper is opened. Total heat value is about the same on an ash-free basis. Lignite and subbituminous have lower heat contents because they contain less carbon.

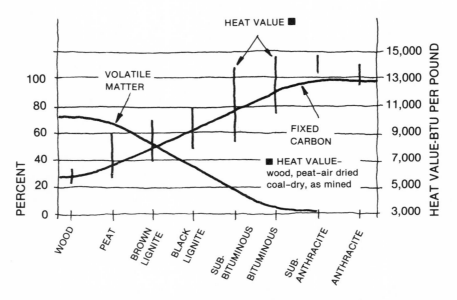

The properties of various fuels. Read heat values on Btu per pound figures at right. Read fixed carbon and volatile matter percentages in figures at left. (Adapted from K. Steiner, *Fuels and Fuel Burners*.)

Weathering of coal needs to be considered with heat value. Just as wood that is left outside deteriorates, coal, when exposed to air either outside or under cover, weathers and loses from 2 to 10 percent of its heat value. Most of the weathering takes place in the first five months after the coal is processed. The process involves the oxidation of the carbon and some loss in volatiles.

Sulfur. This element is found in all coals in various amounts (½ to 5 percent). It generally occurs in one of three forms. *Iron pyrite*, often called fool's gold, is sometimes seen as small specks of yellow in the coal. Upon heating, it becomes iron oxide, with sulfur dioxide (rotten egg smell), or sulfur trioxide, given off. Sulfur dioxide can corrode copper tubes if they are used for domestic water heating in the firebox of the stove. Sulfur trioxide combines with the condensed water vapor in the chimney and forms sulfuric acid. This will deteriorate the chimney in time and combines with rain to form what is known as acid rain.

Sulfur in the *sulfate form* usually occurs as gypsum and remains in the ash. It is present in very small quantities. *Organic sulfur* can occur in varying amounts and is found in the coal tar and coal gas.

Coals having the lowest sulfur content are the most valuable as they meet the Environmental Protection Agency standards for pollution. Mechanical separation and other methods can be used to reduce the sulfur content in other coals.

Buying Your Coal

If you've been heating with electricity, oil, or gas, you've enjoyed the luxury of convenience. Probably you didn't even call to order oil, if you had an attentive dealer. Certainly you did nothing if you relied on gas or electricity. Nothing, that is, but pay the bill, the ever-growing bill.

Buying coal isn't that simple, particularly getting started. You can't just call up your dealer and ask him to bring coal. He'll ask questions, many of them.

Anthracite or bituminous? (He'll probably say soft or hard.)

What size? (Be ready with the size recommended for your stove or furnace, such as nut, pea, or stove.)

How much?

Got a bin for it?

Can I back my truck up so that I can dump into it?

And he'll probably ask some questions about paying for it.

Maybe you should have some questions, too. About price, delivery date, quality of the coal, and whether there are any extra charges, such as for delivery distance.

Here's an explanation of the retail coal business, to help you when you buy that first load of coal.

Delivery to Dealer

Coal is delivered from the mine processing plant either by rail or truck. Your coal dealer, if he is on a railroad siding, will order one or two carloads at a time. These cars contain up to 100 tons. The coal yard generally has bins where the coal can either be dumped or conveyed from the car.

Where rail transportation is not available coal is delivered by truck. A twenty-ton load can be delivered from the mine to the dealer in less than a day in most parts of the United States. Because of the rapid escalation of fuel prices, delivery by truck is more expensive than by rail. This can raise the price charged by the dealer. If you have a choice, it may pay to compare prices. Look in the Yellow Pages for a list of dealers.

You have a choice of bulk or bagged coal, or briquettes. Bulk coal is the least expensive. Dealers will generally deliver quantities of one to three tons at a time. A discount is usually given on quantities of more than one ton at a time. You can save money if you take your own truck or trailer to the coal yard and have it loaded by the dealer.

Bags Cost More

For convenience in handling and for the purchaser of small amounts, coal is bagged in ten-, twenty-five- and fifty-pound units. You will pay extra for the bag and for someone to fill it. Often this amounts to $30 to $40 per ton extra. Some dealers will also allow you to bring your own bags and fill them, using the dealer's scale. Handling coal in bags reduces the amount of dust and provides a convenient unit to move, especially if you live on an upper floor.

Several forms of coal briquettes are available. They are made by compressing fine sizes of bituminous coal into units that are readily handled. Often some type of binder or heat may be used to form the briquettes.

Coal briquettes are similar in form to the sawdust logs or pellets available in department stores. They may be in the log, cube, or pellet form. Before purchasing these units, consider the cost, the amount that you will need during the heating season, and the type of binder, if any, used. They should be used only in fireplaces, or as a starter for a coal fire. Binders such as paraffin, asphalt, or coal tar may create a very hot fire that could damage some stoves.

Cooperative Buying

A popular method of buying coal in some areas is to get together with your coal-burning friends and neighbors and contract for a trailer truck load directly from the mine or through a dealer who has contacts with the mine. This will save 15 to 30 percent, or most of the profit that the dealer normally makes, but it can have some disadvantages. You may have to take delivery at one location or at least within a short distance. The driver won't want to travel all over town to deliver small amounts with a large trailer truck, especially if he has to back into narrow driveways. Second, the quantity you receive may vary depending on the way delivery is arranged. You may get more—or less—than you paid for. Third, the coal may not be as good quality unless you purchase from a reputable supplier.

If you are purchasing coal in bulk you will need a good access for delivery to near where the coal is to be stored. Today things are different from the day when the delivery man would wheelbarrow the coal from the street, across your lawn to a storage bin behind the house. It's also different from the delivery of fuel oil with the truck carrying a 100-foot hose that can be pulled to the tank fill pipe. Have a good access so that the heavy truck can deliver in any type of weather, especially during winter snow or spring thaws. For most delivery trucks the access must be within ten feet of the storage area.

When to Buy

When is the best time to buy your coal? Everyone wants to have the coal bin filled before late fall, so most suppliers are very busy then. Most dealers offer a few dollars a ton discount on coal purchased before the rush season. Purchase yours in July or August and you may get this discount. Also the coal may have less moisture and you will gain a few extra pounds of heat value. If you can afford it and have the storage space, buy your winter's supply at this time.

How Much Do You Need?

Many factors influence the quantity of coal that you will need this winter. We will discuss some of these briefly and then give you a couple of ways to estimate your needs.

Where you live. This is the most important one. If you live in the northern tier of states or Alaska, you know the winters get mighty cold and your fuel usage is large, especially in January and February. If you live in a southern area, you may need heat only a few times a year during the cold snaps.

Size of your house. The heat loss from a building is directly related to the amount of surface area that is exposed; the larger the house the more fuel you need to heat it. Other things like the amount of insulation, the number and tightness of windows and doors, and the windiness of the location are important. Review the section on energy conservation in the Introduction to learn more about what you can do here.

I hope you are not in the situation of the lady who called me recently.

She lived in a 4000-square-foot house using 7000 gallons of fuel oil during the winter. She wanted some ideas on what to do. She explained that it was an older home on a hilltop with a fabulous view in all directions and had many windows to take advantage of this. A lot of energy conservation measures and some structural changes were needed to reduce her fuel usage to an affordable level.

Efficiency of the stove. Just as furnaces vary in efficiency, coal stoves use more or less fuel to give the same warmth. The design of the stove and the way it is installed are important in obtaining maximum heat. These will be covered in the next two chapters.

Operation of the stove. The way you use your stove also influences the number of tons of coal you will use. Some homowners light a stove on the first cold day in the fall and keep it going until it's time to plant the garden in the spring. Others use the stove only on weekends when the family is home.

A few people operate the stove during the fall and don't turn the furnace on until the stove can't supply the needed heat. The furnace is then used as the only heat source during the winter, as it is more efficient the more it runs. Then there are a few individuals whose stove gives off too much heat if operated in the early fall and late spring. They use a furnace during these times to save having to light a fire each day. They often use some wood to supplement the coal. It gives a quick, hot fire to warm the house and then they let the fire go out.

Fuel used. Differences in the heat output of the various types of coal will affect the usage. A difference of 10 percent is not uncommon between two coal beds in the same area. A variation of 30 percent or more can be found between the lowest and highest ranks of bituminous coal. This means that you may need an extra ton of the lower heat value coal to get through the winter.

Estimating usage. Experience is the best method of determining your needs but until you get a winter or two under your belt you may have to rely on another method.

You might use the trial-and-error method. If you are seriously considering using coal as an alternate fuel to heat your home this winter, purchase a good stove and have it properly installed. Buy a ton of coal and operate the stove to see how you and your family adapt to this "new" heating method. If it agrees with you and your coal bin starts running low, order another ton or two. If you find that you can't master the technique or it just involves too much effort, you can always sell the stove and the remaining part of the one ton.

For a more accurate estimation, use the conversion formula based on your present fuel usage (see Table 1) in the Introduction. You will have to decide what percentage of your heat you want to replace with coal and order this amount.

Kindling Needed

You will also need wood to start the fire. The best choice is a mixture of soft- and hardwoods. The softwoods, such as pine, spruce, and hemlock, make good kindling as they split easily and have resins that get the fire started quickly. The hardwoods, oak, ash, hickory, and maple, can be added once the fire is started to give a good bed of charcoal upon which the coal is added. If you can't obtain a mixture, either type of wood can be used by itself. In some areas slabs from a sawmill operation are available. These are a good choice because they dry quickly and split easily.

The wood must be dry. Most woods air dry to a 20 percent moisture content in about a year. This seasoning can be speeded up a little by cutting the wood into stove lengths and splitting it. You can tell if wood is dry by checking for radial cracks in the ends of the logs, and look for bark that is becoming loosened from the wood.

The amount that you need will depend on how you operate your stove, but generally a half a cord should take you through the winter. You will need more if you are going to burn wood rather than coal to take the chill off the house in the early fall and late spring.

Wood can be purchased through many coal dealers, stove shops, local farmers, or wood dealers. If you own a lot of an acre or more that is partially wooded, this may provide all your needs as most woodland will supply at least one-half cord per acre per year on a continuous basis.

Where to Store It

Now that I've told you to purchase your coal during the summer and you've estimated how much you need, it's time to decide where you are going to store it. The best place is probably outside, near the house and under cover. There are several reasons for this choice. First, it keeps coal dust out of the house. Second, it doesn't use up valuable space in the basement, garage, or shed. The location should be near the house, though, so you don't have to track through deep snow during the winter.

Placing the coal bin in an unfinished basement, unused bulkhead, or oversized garage is also quite common.

Coal weighs from fifty to sixty pounds per cubic foot depending on the amount of carbon present and the shape and smoothness of the pieces. Converting this to bulk density means that it takes from thirty-three to forty cubic feet to store a ton. From these figures we can design a bin large enough to hold the coal that needs to be stored. Generally more space is allowed than is needed so that another delivery can be made before the bin is completely empty. Plans for two coal bins are shown.

If you store the coal in a pile on a concrete pad or on the ground, it may freeze and be difficult to use.

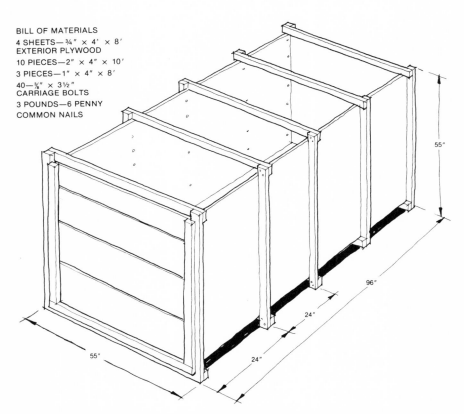

BILL OF MATERIALS
4 SHEETS—¾" × 4' × 8' EXTERIOR PLYWOOD
10 PIECES—2" × 4" × 10'
3 PIECES—1" × 4" × 8'
40—⅜" × 3½" CARRIAGE BOLTS
3 POUNDS—6 PENNY COMMON NAILS

This coal bin, with a 3½-ton capacity, was designed for use in a basement, and should be placed so that coal can be chuted directly into it. A location under a window is ideal. Plywood pieces in front are not nailed in place, but are held by framing, so they can be removed.

Coal as a Fuel

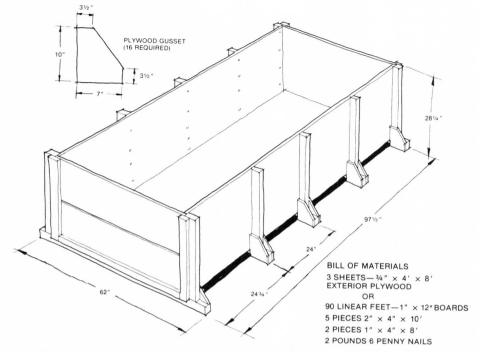

A two-ton-capacity coal bin, suitable for either in the garage or outdoors. Two pieces of plywood on front are not nailed in place, but are held behind 1 × 4-inch frame, so they can be lifted out for easier coal shoveling.

Selecting the Proper Coal

Coal dealers do not stock all types and sizes of coal. They usually stock the sizes that are most often used for domestic heating, namely stove, nut, and pea anthracite coal; nut and stoker bituminous; and pea and stoker lignite. The type of coal most often stocked is that which is the most readily available. If you live in the Northeast, anthracite is commonly available; in the North Central, lignite is found; and in the remainder of the country, bituminous.

Size of Grate

The size of grate openings determines the size of coal to use. Follow the recommendations of the stove manufacturer. Most companies test their stoves using the coal that is available in the area. Stoves that are imported from Europe often are designed to use the size of coal available in the country of manufacture. This generally does not match the size standards in the United States and sometimes problems in operation can result. Several import stoves operate best using a mixture of pea and nut coal.

One other point should be considered and that is the ash content and heat value of the available coal. Often an analysis report is available that allows you to compare one dealer's coal against another. Remember that the higher the heat value and the lower the ash content the more warmth you will get for your fuel dollar. A review of the various coals is made in Table 5. Finally, compare the price of coal among dealers. Often there are significant differences.

TABLE 5. RATINGS FOR FUEL

	Relative Amount Heat	Ease to Ignite	Ash Content	Smoke
Wood	Low	Easy	Very Low	Yes
Lignite	Medium	Easy	High	Yes
Subbituminous	Medium	Medium	High	Some
Bituminous	High	Medium	Medium	Some
Subanthracite	High	Hard	Medium	No
Anthracite	High	Hard	Medium	No
Cannel	High	Very Easy	High	Some
Coke	Medium	Hard	Medium	No

How Coal Burns

It's important that we understand how coal burns so that we can select the right stove and then operate it to obtain maximum heat.

Combustion

Heat is developed from the chemical change brought about by the combination of the combustible parts of the coal with the oxygen in the air. Carbon dioxide, carbon monoxide, and water are formed. In fuels having a moderate amount of volatiles, smoke can also be formed. This is an indication of incomplete combustion.

This occurs when one or more of the following conditions exists:

1. Air is insufficient.
2. Poor contact between the fuel and the air.
3. Incomplete mixing of the air and gases.

Incomplete combustion means that fuel is being wasted and the efficiency of the burning is being reduced.

An efficiency of 100 percent is never reached. Coal stoves operate in the 50 to 70 percent range with lower efficiencies occurring during start-up. Heat is also lost when the moisture in the coal is turned to steam and escapes up the chimney.

Another large chunk of heat is lost warming the extra air that passes through the firebox. Air is made up of 79 percent nitrogen and 21 percent oxygen by volume. Only the oxygen is used in combustion. The inert nitrogen is a detriment because it robs heat from the fire as it travels through the firebox.

Let's look at the burning process another way. The first step is to heat the fuel to drive off any moisture present. With wood and the lower forms of lignite and subbituminous coal it takes a considerable amount of heat to boil off this water. Anthracite and bituminous coals have much less moisture.

Once this moisture is removed, ignition can take place. For wood this occurs at approximately 550° F., for coal between 750° and 950° F. (Table 6). From this you can see that a coal fire is harder to start than a wood fire.

After ignition the fuel gets hotter and the volatile gases are released. If the temperature in the firebox is hot enough and adequate oxygen is present, these gases will burn, releasing the heat. If not, the gases are carried up the chimney unburned.

In wood about 50 percent of the heat is in the volatile gases. Because of

TABLE 6. IGNITION TEMPERATURE OF COMMON FUELS

Fuel	Ignition Temperature
Paper	350°
Peat	435°
Wood	550°
Fuel oil	560°
Lignite	600°
High-volatile bituminous coal	765°
Low-volatile bituminous coal	870°
Anthracite coal	925°

this, a large amount of air, called secondary air, is needed in the area above the fire. A wood stove is generally designed so that approximately 80 percent of the air entering the firebox is secondary air. For lignite that has a lower volatile content, less air is needed as secondary air.

Bituminous coal contains 30 to 35 percent volatiles and about half the air is needed above the fire and half below. Anthracite, on the other hand, is mostly carbon with only 3 to 8 percent volatiles. Stoves designed for this coal should have only 20 percent secondary air.

The final step in the burning process is to recover the heat from the coke, which is what remains when the volatiles have burned off. It burns in the range of 1500° to 2500° F. This is why coal stoves are designed with high-temperature firebrick or cast iron as a firebox liner. These temperatures would warp or burn out a mild steel firebox in a short time.

Primary Air

The air needed to supply oxygen in the fuel area is called primary air. It enters low in the firebox on a wood stove and travels up through the grate in a coal stove. For wood, only 20 percent of the total air should enter in this area. For coal, on the other hand, up to 80 percent should be primary air.

As you can see, there are significant differences in the way that coal and wood burn. It is important, then, that if you are selecting coal as a fuel or if you may at some time be burning coal, you purchase a stove that has been designed to burn coal.

CHAPTER 2
SELECTING A STOVE

With thousands of models of stoves on the market, it can be difficult to make an intelligent choice. In this chapter we will try to sort out the many factors that are important and those that are not so important. One of the questions that I am often asked is what is the best stove? There is no one best stove for all uses and locations, even though some stove manufacturers and dealers would like you to think so.

Don't rush out and buy the first stove you see. Many people whom I know have changed stoves several times before they got the one that they like. Carefully plan how you want to use the stove and also how it will best fit into the area where it will be placed. List those features that you must have in the stove, those that you would like, and those that are optional. The checklist at the end of this chapter can be used as a guide.

What Use?

The most common reason for installing a stove is to supplement the existing heating system and reduce fuel usage. To some people this means operating the stove during the long winter evenings when you are home. To others it means burning a fire continuously to heat the living areas. In both cases the savings comes from not having the furnace run as much. The stove could be one of medium size, having an appearance that fits the decor of the room, and connected to the fireplace flue.

For folks who live in areas where snow and ice storms are frequent and power outages last several days, a coal stove can be their means of survival. It means available heat during these periods for comfort, for cooking, and to keep the pipes from freezing. The stove for this purpose may take on a different look, having a flat surface for heating pots, and plain appearance, as it may be located in the basement.

Some people, once they have had a year or two of experience with a stove, decide that they like the idea and install a second and sometimes a third stove to heat different parts of the house. The even heat of a coal fire and the fact that you can move closer or farther away as your body comfort demands are great advantages. The savings in oil or gas from heating only those rooms that you are using at the time will often pay for the stove in a short time. Another advantage of the multi-stove installation is that you can operate only those stoves that are needed to heat the house, especially in the spring and fall when just the chill needs to be removed.

Status Symbol

In many homes being built today the stove has become a status symbol, just as the inefficient fireplace was a few years ago. When these homes are built the chimney flue is included but the choice of stove is left up to the new homeowners. If the stove is to be used as the focal point or conversation piece, one of the more ornate European designs with fancy castings and chrome trim may be the best choice.

In some homes the ideal stove may be a coal-fired kitchen range. It will heat the kitchen area and provide plenty of space for simmering that pot of stew or baking the week's supply of bread. Many cookstoves have the option of attaching a hot water pre-heater, too.

The coal-fired range is also a good choice for those living in smaller apartments, condominiums, and vacation cottages. Its heat output is often adequate for all but the coldest days. The range can also be taken with you when you move to a new location.

Location

Next you have to decide where it can be placed. Here the decision revolves around where space is available, where the chimney is or can be located, and where the stove will do the best job of heating.

The amount of space that is needed for the stove is determined by its size and shape. For example, often we look at the family room and say a stove would look nice against one wall. We may forget, though, that the minimum distance from the stove to combustible walls and furniture is three feet without special protection and this would make the stove project halfway across the room. Often the stove fits best along one of the center walls in the house. This is where the chimney is located or where a factory chimney can be placed.

A central location in the house is better for heat distribution but it's often more difficult to locate a chimney here, especially if you have a

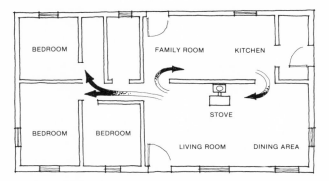

Here's a good location for a stove in a ranch house. Heat will circulate in the living areas, and bedrooms will remain cooler.

Cape Cod or raised ranch. A chimney running through an upstairs bedroom is unsightly.

Locating the stove in an existing fireplace can solve the chimney problem but the majority of fireplaces are located on the end wall making heat distribution more difficult. Many manufacturers take advantage of this and have developed stoves that either fit into the fireplace or on the hearth in front of it and use a blower for circulation. A further advantage to this location is that the brick or stone in the fireplace is a large heat sink which continues to warm the room long after the fire has gone out.

Place It in Basement

The best location in some homes is in the basement. The full benefit of the heat is then used as it travels up from floor to floor. A better draft from the higher chimney and the location of the stove and its associated dust are further advantages. With this location, though, some heat is lost to the basement walls. Heat can be circulated by using the basement stairwell and several floor registers at points farthest from the stairs.

Several manufacturers have developed stoves with blowers that will move the heat through ducts to all sections of the home. A coal-fired furnace could also be used in the basement. Often it is possible to take advantage of the distribution system of the oil or gas furnace by using an add-on unit.

Size

The next step is to determine what size your stove should be. Often a stove is installed that is too large for the area to be heated. It is better to size it to heat the area in the milder part of winter than for the coldest day of the year. Although most stoves have a fairly wide range of heat output, having too large a stove can make the house uncomfortably warm during the spring and fall.

There is no general standardized testing method for rating stove output. This makes it difficult to compare stoves. Each manufacturer has his own method. Here are three ratings you may see. You also have to take into account that your fuel may be different.

1. *Area heated.* This is the number of rooms or square feet of floor area that the stove will heat. To calculate the area, multiply the width and length of each of the rooms to be heated and add the areas.

2. *Volume heated.* This is area heated multiplied by eight, the height of the room. Both this value and area heated assume some level of insulation and tightness to the home. If you have an older home with little or no insulation and you can feel the cold draft on your feet during the winter it would be best to use about two-thirds of the figure stated by the manufacturer. On the other hand, if yours is a modern, tight, well-insulated home, you can probably add a third to the value and still have the stove do a good job.

3. *Btu output.* Unless you are an engineer or have a friend who is, this value will have little meaning. To use this rating you have to be able to calculate the amount of heat that is escaping each hour from the different areas of your home. It also requires a knowledge of the amount and kind of insulation present. The Btu output can be used, however, to compare two stoves. The stove with the higher output will provide more heat per hour during the burn cycle.

The two factors that control the heat output of a stove are the area of the grate and the draft in the chimney. This assumes that adequate air can get to the fire. The larger the grate area and the taller the chimney the more heat a stove will produce.

How do you tell if the stove you select is safe? A stove can fit into one of three categories.

1. *Listed.* This means that the stove has passed strict testing procedures done by an accepted testing laboratory. The procedure consists of placing the stove in a room in which temperatures can be recorded in more than 50 locations on and around the stove. Several types of burn and smoke tests are conducted including a flash fire test in which the drafts are left wide open. The instruction manual on how to install and operate the stove is followed exactly. If the stove passes these tests it is listed. A regular check is made of these stoves during and after manufacture to insure that

the same quality and workmanship are in all the stoves. You can be reasonably sure that if you have a *listed* stove it is safe.

2. *Approved.* These are stoves that are acceptable to a local or state building inspector or fire marshal. Tests may have been conducted on the stove or it may just have been inspected.

3. *Nonlisted or nonapproved.* Because of the high cost of getting a stove design tested (several thousand dollars) many small manufacturers have not taken this important step. Most of these stoves are safe and would pass the test with slight modifications. Still, the building inspector may not let you make the installation.

Several states with state-wide building codes require that a stove be *listed* before it will be approved by the building inspector. Check with your local authority before purchasing the stove.

Radiant Stoves

Stoves can be classified as either radiating or circulating. It is important that you recognize the difference as the safe distance of the stove to combustible materials can be much less with the circulating type.

Radiant stove

Radiant stoves heat objects and people by energy that is transmitted in the form of waves just as we are heated by standing in the sunlight. The heat from the fire is conducted through the firebrick, steel, or cast iron and transferred to the room by radiation waves and convection currents. Most radiant stoves are painted black because this color is the best for radiation.

When facing a radiant stove the front of your body feels warmer than the back. Also, the closer you get to the stove the warmer you feel. This is because you are absorbing more of the energy waves. Another advantage of a radiant stove installation is that if you are near the stove you feel comfortable even if the air temperature in the room is near 60° F.

A disadvantage to radiant-type stoves is that unless the stove is very heavy the heat output will be quite varied over the burn cycle.

Circulating Stoves

If we place a sheet metal enclosure around a radiant stove, allowing the cool air to enter under the enclosure, be heated by the stove, and exhausted out the top, we have a circulating stove. This stove works on the principle that as air is heated it expands and rises. The air currents that

Circulating stove, showing air entering at base, moving up through the cabinet, and, heated, emerging from the top.

are developed circulate the heat throughout the room. For this reason it is a better choice for larger rooms.

The two- to four-inch air space left between the stove and the enclosure keeps the surface of the enclosure much cooler. This allows the stove to be placed closer to combustible walls and is safer when young children are part of the family.

Some manufacturers make a forced circulating stove. Ducts, pipes, or channels are built into the stove and a blower is attached to give a positive air movement. The better systems use a thermostatically controlled fan which turns off when the fire dies down. Otherwise, cold air will continue to circulate in the room, giving a chilling effect. This type of stove is slightly more efficient in its use of coal as the heat transfer surfaces are kept cooler and absorb more heat from the fire.

COAL IN A WOOD STOVE?

"You can burn wood in a coal stove but not coal in a wood stove."
This is an old saying, and it still holds true today.

As explained in the chapter on coal, wood and the higher ranks of coal have different burning characteristics. Coal requires more primary air below the fire, with the air moving up through the bed of coals. Coal burns hotter than wood.

You should not attempt to burn coal in a stove without a grate. You would be wasting your time—and it could be dangerous.

Some stoves are labeled "For Wood Only" or "Not to be Operated with Coal." There are several reasons for this:

1. The unit will not operate safely with coal; the stove bottom or sides lacking protection for the higher temperatures developed, may burn out.
2. The stove has not passed a certified test using coal.
3. The manufacturer knows the stove will not burn coal well, and doesn't want his reputation damaged by unsuccessful attempts with coal.

Some wood-stove manufacturers are attempting to modify their stoves to burn coal by casting or fabricating a basket grate to fit in the bottom of the stove. This basket usually does not have a shaker, making it difficult to remove the ashes. The amount of coal the basket grate holds is small and its depth is shallow, so the length of burn may be hard to regulate. While some of these units may be passable for occasional use, they are not acceptable if you are planning to make a permanent switch to coal.

Some stoves are sold as wood-coal, coal-wood, or solid fuel combinations. The fuel that the manufacturer emphasizes in his literature will generally indicate which fuel will burn best in the stove.

Materials

In your visit to the local stove shop you will find that many materials are used in the manufacture of stoves. The main body of the stove is either cast iron or steel.

Most of the best coal stoves are made from cast iron. This material is iron combined with carbon and silicon. The parts for the stove are cast from molten metal poured into sand molds. After cooling, the castings are cleaned and machined to fit tightly together. Asbestos rope, furnace cement, or high-temperature fiberglass is used to seal the joints. A check of the castings in the stove for thin spots, large sand pits, tight joints, and warping will indicate the quality of the stove. Be especially careful in purchasing stoves made in the Far East as the cast iron is often imperfect and the workmanship poor. I have seen some of these stoves with holes at the joints large enough to stick a finger through.

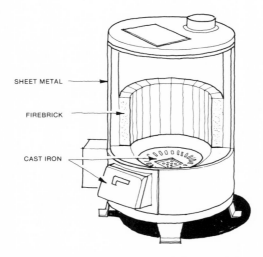

Stove materials include cast iron, firebrick, and sheet metal.

Cast Iron

Cast iron is used because it can take the higher temperature of a coal fire without oxidizing or burning out. Many cast-iron stoves have been in use for a hundred years or more. In some stoves cast iron is used only for the door and stovepipe connector because it does not warp as easily as steel.

One disadvantage to cast iron is that it is more brittle than steel. Sudden temperature changes such as opening the door rapidly and allowing a lot of cold air in may stress the metal so that a crack could develop—and it is almost impossible to stop a crack from lengthening. Expansion and contraction from heating and cooling keep it working. Drilling a small hole at the end of the crack or having a good welder braze the joint may work for a while.

Most cast-iron stoves must be taken apart every few years to replace the seals at the joints. The stove bolts holding the stove together may have to be chiseled off and replaced. The joints should be scraped and sanded lightly to remove the old joint compound and a new layer of furnace cement applied before reassembly. Furnace cement can be purchased in pint- or quart-size cans at most hardware stores or stove shops. It should be allowed to dry at least a day before the stove is used.

Cast iron is also used for the grates and firebox liner in coal stoves. These are made so that they can be easily replaced if they burn out or crack. The larger stove shops carry replacement castings for the stoves they sell.

Steel

Steel is used in the construction of some coal stoves and many wood stoves. Its low price, combined with great strength, allows its use in the parts of the stove that receive the greatest stress. Steel can be formed into complex shapes and into large pieces to eliminate many of the joints found in cast-iron stoves. It can also be welded easily.

For material less than 3/16 inch thick, steel is designated by a gauge number. The lower the number the thicker the sheet steel. This material is used in some low-cost wood stoves and as the outside shell in some firebrick-lined coal stoves. Plate steel is thicker material and is often used in the form of boiler plate or diamond plate.

Thinner steels are more apt to warp from the heat of the fire. This can affect the fitting of doors and can stress welds until they eventually break. The high temperatures developed in the firebox will cause oxidation of the steel, hastening its deterioration unless protected by firebrick.

Both cast iron and steel have the same heat transfer characteristics. The rate of heat transfer depends on the material thickness; that's why a thin sheet metal stove tends to give heat more quickly. It also tends to cool off faster when the fire goes out.

Important Features

Liners

Many coal stoves made today have firebox liners made of firebrick, fire clay, or ceramic. This gives a stove several advantages over an unlined one.

1. It adds weight or mass to the stove, increasing its heat storage capacity. Although the stove heats up more slowly when started, it gives off heat for a longer time when the fire goes out. Some stoves have as much as 200 pounds of brick.

2. The firebrick protects the metal in the hot firebox area, extending the life of the stove.

3. The liner helps to retain a more uniform fire and, therefore, more uniform heat output.

Firebrick is made from a clay containing silica, alumina, and iron. It can take temperatures of over 3000° F. Sometimes firebrick is placed in the stove loose and held in place by metal retainers or brackets. This allows the brick to expand when heated. Other manufacturers cement their brick in place. This way the ash from the fire doesn't fill the cracks. A few stoves have a liner cast in the shape of the firebox.

With hard and constant use, most firebricks break down or crack over a period of years. The brick can be replaced with material available at most stove shops. The cast liner is much more difficult to replace because the stove must be disassembled. If the manufacturer has changed design the liners may no longer be available. Sometimes individual firebrick can be fitted and cemented in place. This may require the services of a mason.

Most coal stoves feature airtight construction. This allows better control of the draft and a more uniform fire. To gain airtightness the joints must be sealed and the doors and other openings must close tightly.

Tight Door

Several methods are used to ensure a tight door. Some European manufacturers use castings with machined surfaces that fit tightly together. The most common method is to cast the door with a groove around the perimeter. Asbestos, fiberglass, or other nonmelting material is pressed into the groove allowing some excess to show. When the door is closed a seal is formed between this material and the door frame. Replacement materials are found in most stove shops.

Door size and location are important in fueling the stove and removing ashes. Most coal stoves have an ash door and one or two fueling doors. The ash door located below the grate area allows the ash pan to be removed and emptied. Except in stoves with thermostats it also contains the primary air draft control. When purchasing a stove, check to see that this door and the draft control close tightly. If not, it may be hard to maintain a uniform fire.

The fire door above the grate is used for adding the coal. It should be large enough to allow a shovel or the end of a coal hod to be inserted. The door should be protected by a perforated plate inside. This cools the door to keep it from warping and allows the secondary air to be preheated before it is mixed with the volatile gases.

Windows

Fifty years ago when coal stoves were popular a window was placed in the door so that you could watch the fire. Several manufacturers are now adding this feature to today's stoves. Tempered glass, pyrex and isinglass are often used. Isinglass is thin, semi-transparent layers of mica that are not affected by the heat.

The fire door should be high enough above the grate to give five to ten inches of depth of coal. In firing the stove, the coal is maintained at the level of the bottom of the door. On some stoves the fire door or lid is placed on the top of the stove. Although you don't have to bend to add the coal there is a greater chance that smoke and fumes will enter the room when the door is opened. This can be minimized, though, by opening the bottom draft for a few minutes before adding fuel.

On some circulating stoves the top lifts out of the way to expose the top of the stove. This top can be used for a cooking surface or to place a pot of water to increase humidity. Most circulators also have a large access door in the enclosure to expose the ash and fire doors.

Draft Regulators

The rate of burn in a stove is controlled by the draft regulators. These can be manually operated slides, screw caps, or hinged flaps. They can also be thermostatically operated closures.

These have to be designed to operate under different conditions depending on the fuel used. When starting the fire with wood or when burning anthracite or coke, three-fourths or more of the air should be supplied through the bottom draft. If you are using wood, lignite, or high volatile bituminous, one-half or more of the air should enter through the top draft. These air inlets should be large enough so that they don't have to be fully opened to get the fire to burn.

Thermostat

A few manufacturers offer a thermostatically controlled stove. The thermostat senses the temperature of the air near the stove and adjusts a damper to allow more or less air into the stove.

The sensing element can be either a bimetallic strip (two metals with different expansion rates) or a thin wafer that contains a liquid. This element should be placed where it is not affected by drafts and air movement in the room.

As the thermostat opens and closes it controls the position of the damper. On coal stoves a two-stage damper may be used. As the thermostat calls for more heat a small hole may be opened, allowing some primary air to enter. If the room cools off more or you desire more heat a second larger damper may be opened.

To provide for different levels of primary and secondary air, top and bottom dampers may be tied together by a rod and activated by the one sensing element. In some stoves primary and secondary air enter through one inlet and are channeled within the stove. The proportioning is done by the size of slots, channels, or holes, and is fixed. These stoves work best for the type of coal that the stove was designed for.

The advantage of a thermostatically controlled stove is that it will

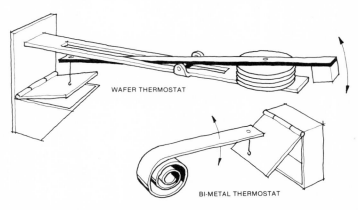

In stove thermostats, the sensing element is either a bimetallic strip, containing two metals with different expansion rates, or a thin wafer that contains a liquid.

Selecting a Stove **49**

maintain a constant heat output within reasonable limits. It is a good choice if the stove is in the basement as it may require less attention. Those away from home all day will appreciate the thermostat because it can adjust for varying weather conditions that may affect the burn rate.

As with other mechanical devices, thermostatically controlled dampers occasionally stick or bind. Cleaning to remove dust and grime and lubricating the moving parts with graphite or a high temperature oil will help.

Firebox (Firepot)

Buying a stove with the proper firebox size and shape is more important with a coal stove than a wood stove. Because coal burns in different layers, a firebox capable of getting the depth of coal needed will give an even burn cycle. A taller firebox also allows enough coal to be loaded at one time to get the fire to burn for many hours. Depending on the chimney draft and the damper openings, most domestic stoves will burn coal at the rate of ½ to 1½ inches per hour. In a good coal stove you should be able to get at least six inches of coal above the grate.

A second factor that affects the operation and heat output of the stove is the ratio of the grate area to stove surface area. If the surface area is not large enough, heat will be wasted up the chimney because it is not absorbed by the walls of the stove and radiated into the room. On the other hand, where the wall surface is too large in relation to the grate, too much heat will be absorbed from the fire and the flue gas temperature may drop too low to keep up a good draft. Ideally, the gases should be in contact with the stove until they are cooled to about 600° F.

Some manufacturers increase the surface area of the firebox by adding flutes, corrugations, or baffles. These can be cast into the sidewalls or can be a separate channel through which the flue gases must travel before they enter the stovepipe. The flutes or channels also help in some stoves to get extra air up through the coal.

Grate

The grate is one of the most important parts of a coal stove. Most wood stoves operate well without a grate, but the coal stove needs one for several important functions.

1. It supports the ash layer and the fuel above it.

2. It allows the removal of some of the ash layer without disturbing the fire.

3. It allows the primary air to be distributed evenly to the fuel layer and to pass up evenly around the pieces of burning coal.

4. It can be used to settle the coal down if it starts to cake or wedge itself into the firebox.

Almost all grates are made of cast iron. The quality of the casting and the care with which the stove is operated will determine how long they will last. A grate can be warped or burned by shaking the fire too much so that the hot burning coal rests on it. Most stoves are designed so that a damaged grate can be replaced. Some stove dealers stock replacement grates.

Grate design can affect the way a stove operates. I have received many calls from stove owners having problems in keeping a fire operating for more than a day at a time. A poor grate design is often the cause. A good grate will have the following features:

1. A system for shaking the whole grate area. Some grates have no shaker. This means that you have to insert a poker or "fiddle stick" to get rid of the ashes. This mixes the ash layer and the molten coal layer above it and forms clinkers. These clinkers then block the grate openings and restrict air flow.

Types of grates used in coal stoves

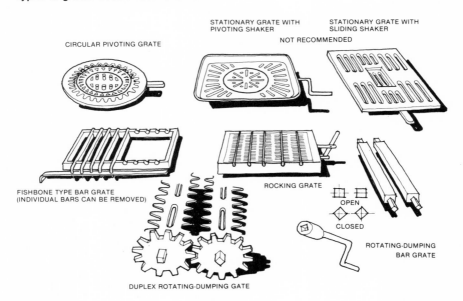

Some grates have only a movable center section. Ashes tend to build in the corners and again restrict air flow. Often this stifles the fire and causes it to go out. It's a nuisance to have to remove the coal and start a fire every other day.

In a whole grate shaker system the grate can rotate around a center pivot, rotate in sections, or slide back and forth. An added feature with some grates is that they can dump the ashes and unburned coal into the ash pan. This is very convenient in the spring and fall when you may not want to operate the stove continuously.

2. The spaces between the bars will be sized to support the size of coal that the stove was designed for. Each manufacturer designs the grate for the coal marketed in the area where he expects to sell his stoves. This stove may not work well if this coal is not available. For example, some European coal stoves work best with a mixture of pea and nut anthracite coal. This mixture is not available in the United States.

3. The air space in the grate will be one-third to one-half of the grate area and should be evenly spaced. This allows for support of the ashes and still allows good air movement.

4. The bars are of a tapered design so that the ashes will not wedge between them.

5. A solid system supports the grate so that it can carry the weight of the coal without stress.

Ash Pit

The stove should have a large ash area preferably with an ash pan located below the grate. If you burn forty to fifty pounds of coal a day, the average usage for a stove, you will get almost a gallon of ashes. If the stove has a small ash pan, you will have to empty it each day. This leads to dust and ashes that have to be cleaned off the floor or stove pad.

The level of ash in the ash pit should be kept well below the bottom of the grate. This will insure that adequate air reaches all parts of the grate, and will prevent the grate from becoming overheated.

Appearance

Most manufacturers try to design each stove so that it is appropriate to the location in the home where it is placed. For the living room-family room area, ornate designs with castings depicting a woods or animal scene in relief carving are common. These designs have a larger radiating surface than a flat plate or cast stove.

The beautiful porcelain enamel finish on some stoves is available in several colors. This finish is created by applying several coats of powdered glass and baking it at a high temperature. Porcelain enamel coatings have excellent corrosion resistance but can be brittle and chip, especially at higher temperatures. They also add $100 or more to the cost.

Most stoves are coated with a high-temperature black paint. This paint adheres well and lasts many years but may need to be touched up from time to time. Spray cans are available at most stove shops. Stove polish can also be used to maintain a cast-iron stove. It is available in liquid or paste and is applied with a piece of cloth or brush.

To dress up a stove, manufacturers often use chrome or chrome-plated steel strips. Nameplates, door trim, controls, and caution labels are treated in this manner.

Stove Efficiency

Is one stove more efficient than another? This question is often asked as the prospective purchaser views twenty or thirty models and sizes of stoves at a local stove shop.

There are many types of efficiencies that can be applied to coal stoves but the one that has the greatest meaning for us is the Energy Efficiency. It is defined as the percentage of energy in the coal that is converted to useful heat in the home. It includes the heat received from the stove, the stovepipe, and chimney.

Energy efficiency of a coal stove installation depends on the design of the stove, how it is installed, and the skill of the operator. Let's look briefly at these.

To get high efficiency in the burning of the coal the proper amount of air must be mixed with the volatile gases. This means that the primary and secondary air supplies must be sized properly so that good mixing occurs in all parts of the stove. The temperature in the various zones of the burning coal must also be maintained so that complete combustion takes place. Firebrick and cast-iron liners help by retaining part of the heat of the fire.

Heating Air

Preheating the air entering the stove, if it is warmed several hundred degrees, can make a stove more efficient. This is because less of the heat in the firebox is needed to heat the air and a warmer temperature is maintained. The air leaks that a stove has are another factor in stove design. The more air leaks there are, the lower the firebox temperature and the lower the efficiency.

A stove can be efficient in the combustion phase but inefficient in conducting the heat away from the fire. The surface area must be large enough to radiate the heat from the stove. The short rectangular stoves and converted box stoves generally do not have enough radiating area for a higher rate of fire.

Remove Soot, Fly Ash

Maintenance of the stove is also important. Radiating surfaces, ledges above the firebrick lining, and flue gas passages should be kept free of fly ash and soot, both excellent but unwanted insulating materials.

A longer stovepipe can help recover heat not radiated from the stove. Three lengths of stovepipe and an elbow have as much radiating area as some of the smaller stoves. Locating the stove where free circulation can take place—not tucked into a corner or in a fireplace—will increase efficiency. It is also possible to increase efficiency with a stove that uses a blower to circulate air past the radiating surfaces.

A stove must be operated properly to get maximum efficiency. Adjusting the draft regulators for the different parts of the burn cycle and firing at the optimum rate for a particular stove will help.

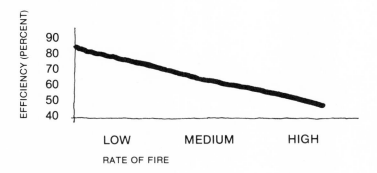

Efficiency of coal stove based on rate of fire

All of the above factors will change with the firing rate of the stove. The accompanying diagram shows the efficiency chart of a typical coal stove. Maximum efficiency is attained at a very low firing rate.

Efficiency Tests

In one test, a well-insulated room is used to measure efficiency. The stove is set up and fired in the room. Air is circulated through the room and the flow rate plus the temperature difference between the air entering and the air leaving is measured. Figuring with the number of pounds of coal used and the time period, the efficiency can be calculated.

The Bacharach test is a second method used to measure efficiency of oil and gas furnaces. Flue gas temperature and carbon dioxide measurements are taken and the efficiency level is obtained from a chart or table. This method can be used with coal if a chart can be obtained for the type of coal you are using.

A third technique that will give a reasonably accurate picture of the energy efficiency of your stove is to measure the reduction in the amount of oil or gas used to heat your home, after coal supplies some of the heat. Two factors that can affect this greatly are:

1. The difference in the weather between the time when you are using coal and the comparable period when you used only oil or gas.

2. The efficiency of the oil or gas furnace.

As you can see, it is difficult to get a meaningful efficiency rating for a stove. Most coal stoves fall within the 50 to 65 percent range for day-in, day-out operation. Be wary of manufacturers or dealers who report efficiencies in the 80 to 90 percent range. These, if they are accurate, were obtained under conditions that you can't duplicate in your home.

Heating Part of Your Home

A first step for many homeowners is to install a unit that will heat part of the home, usually the living room-family room or dining room.

Fireplace. The least efficient method—5 to 15 percent efficiency at best—is to place a coal grate in a fireplace. The large opening and the poor control of the damper allow most of the heat to escape up the chimney.

During the spring and fall a coal fire can be used to take the chill off the room but during the winter in the colder climates you will probably end up with a net loss. Many people find this out the hard way during extended power outages. The room having the fireplace fire burning is the

Selecting a Stove   55

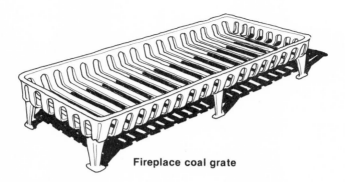
Fireplace coal grate

coldest one in the house. Eventually they move out of the house to a neighbor's that is nice and warm, heated by a coal stove.

For the occasional fire, a fireplace coal grate with cannel coal or coal briquettes works well. Five to ten pounds placed on a bed of hot coals will burn for several hours. If you want the fire to burn through the night you will have to get up and reload it two or three times. If the fire goes out, the chimney will continue to draw heated air out of the house all night.

Fireplace insert. The installation of a fireplace insert makes more sense. Many manufacturers have seen the potential for this application and are making inserts that burn coal. These are designed to fit the more common sizes of openings and have a plate that seals the front of the open-

Fireplace insert

ing. Some are made to fit flush with the face of the fireplace; others extend onto the hearth.

For more efficient operation look for an insert with double-wall construction and a thermostatically controlled blower. These help to circulate heat in the room. Firebrick and cast iron are better than steel for the firebox, and a shaker grate is a must for ease of operation. To retain the view of the fire a tempered glass insert is often used. Many of these units will burn eight to ten hours on one loading of coal and will heat two or three rooms.

The fireplace unit must be installed according to the manufacturer's instructions. A tight seal must be maintained around the trim panels to get adequate draft.

Franklin stove. The Franklin stove was not designed to burn coal; however, it is possible to insert a grate that will hold a coal fire. Again, soft coal or briquettes will work better as they ignite and burn more easily. A large fire should not be built because most Franklins do not have added firebox protection for the higher temperature developed. Also, most of them are not airtight and draft control may be poor. Use of coal in a Franklin stove can be dangerous.

Parlor stoves. These free-standing stoves will fit the decor of almost any room. Most manufacturers make several styles and sizes that are designed to burn coal. For best results and ease of operation, look for the following features:

1. Cast-iron or steel-plate construction
2. Cast-iron shaker grate
3. Airtight construction
4. By-pass draft system
5. Easy loading door
6. Large ash pan
7. Cook top

Most parlor stoves have the stovepipe connection in the rear. This makes it convenient to place it in front of the fireplace or on the hearth and connect into the fireplace flue. These units also work well when attached to a masonry or factory-built chimney.

Several manufacturers offer a circulating design with a porcelain enameled cabinet. Others use enameled cast iron. Mica or tempered glass view doors are common on many models.

Parlor stove

Most parlor stoves will hold thirty to fifty pounds of coal and give a burn time of twelve to twenty-four hours on one loading, making heat output even and operation easy. Before purchasing the stove check to see what type and size of coal is required, as several European designs work best on a combination of pea and nut anthracite, not sold in this country.

Upright stoves. This design for a stove lends itself best to the way in which coal burns. The height of the firebox is usually greater than either the width or length. A small grate area supports the coal and as it burns it self-feeds.

Many stoves of this style use a thin, blue or black steel sheet to enclose a firebrick liner. Cast iron is used for the base, the top, and the doors. The top is usually flat and can be used for cooking or to hold a pot for adding moisture to the room.

Upright stoves usually cannot be connected through a fireplace opening because the chimney collar is located too high. They are good where you have limited space and can't put a stove that will stick too far into the room.

58 HEATING WITH COAL

Some manufacturers make several sizes, the largest holding over 100 pounds of coal and capable of heating several rooms even in the coldest weather. Draft control usually consists of vents in the ash door and near the top of the stove.

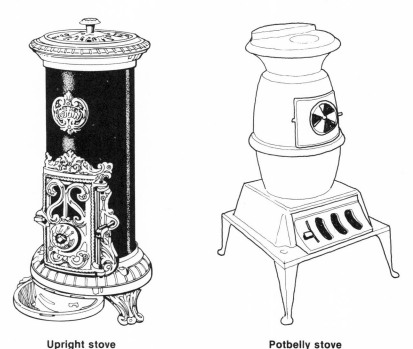

Upright stove **Potbelly stove**

Potbelly stoves. If you want to have the nostalgia of the pioneer days, the potbelly is a good choice. These stoves occupied the place of honor in the railroad station and country store where they could distribute their warmth to all corners of the building. The chrome footrest helped warm cold feet on a winter day, and drying racks held wet gloves.

These stoves are still available and burn well using coal. One word of caution: purchase only well-made American stoves as the Far East imports often have poor quality castings that may crack, warp, or burn out quickly with constant use.

Potbellies are generally not airtight in design because the doors and cooking covers are loose-fitting, and so they are less efficient. It's important that a damper be used in the first section of stovepipe to help give better control of the fire.

Heating Most of Your Home

Where large amounts of heat are your objective, look to the coal heater, which is generally placed in the basement. These thermostatically controlled units with heavy firebrick lining have as much capacity as many home furnaces. Provisions must be made, though, to distribute the heat throughout the house. The use of the basement door, floor and ceiling registers, and small fans is common. Before making a choice of a heater, read Chapter 5 to see whether a furnace or boiler might be better.

Pay particular attention to the grate system in selecting a heater. These units will hold more than 100 pounds of coal and if the fire goes out the unburned coal must be removed to get the fire started again. The grate should shake the whole fire area and preferably be able to dump the ash and unburned coal into the ash pan.

Most heaters are available in either radiant or circulating design. The circulating design is a better choice for large areas, such as an open basement, as it will maintain a more uniform temperature. A few units have an attachment to preheat domestic hot water.

Coal heater

Cookstoves. For many of us the use of coal goes back to the ornate kitchen range our grandmothers used to cook the delicious holiday dinners. These stoves with the large baking and warming ovens, copper water tank, and drying racks are still available from several manufacturers.

Although these stoves are best suited to cooking and baking they will heat several rooms in the home. These stoves are better suited for heating using coal rather than wood because the small firebox does not hold much wood and it must be cut into short pieces. The double-draft system and cast-iron shaker grate make the stove easy to operate.

The more modern kitchen ranges use the same basic firebox design but the stove is enclosed in an enameled steel outer shell. This allows the stove

Old-fashioned cookstove

Modern coal cookstove

to be placed near the wall and cabinets. Some stoves are available with a pipe coil to preheat domestic water.

One disadvantage to the kitchen range is that the fire has to be kept going continuously or the homemaker has to plan ahead several hours so the stove is hot when it's time to start the meal. Several ranges are available with dual fuels—coal and gas or coal and electric. These are convenient in the summer when the heat is needed only for cooking.

Used Stoves

What about purchasing a used stove? We see them offered every day in the classified section of newspapers and at auctions. Some stove dealers also offer them for sale. Let's look at why these stoves are on the market.

Homeowners' needs change. The present stove may be too small or too large or it may not fit the decor of the room. Most of these stoves have been used only a year or two and are good buys.

Some people have had it with burning coal and wood. It's just too much trouble having to tend the fire, clean the ashes, and lug in the coal. Again, a good buy if the stove is fairly new.

Older stoves may have had hard use. Check for warped or missing grates, cracked or poor fitting castings, and broken hinges and poor construction. If the stove is fairly new, parts are often available and the stove can be repaired. If it's old and the manufacturer is no longer in business it may be impossible to get the stove back in operating condition. Several companies still carry parts for some of the older stoves. These include:

Aetna Stove Company, Inc.
SE Corner, 2nd & Arch Streets
Philadelphia, PA 19106

Avondale Stove & Foundry Co.
2820 6th Ave. S.
Birmingham, AL 35233

Brewer Atlantic Clarion Stove Co.
Brewer, ME 04412

Michael Brucher, Inc.
501 Homan Ave. N.
Chicago, IL 60624

Charleston Foundry Co.
1616 Pennsylvania Ave.
Charleston, WV 25307

Empire Furnace & Stove Repair
793 Broadway
Albany, NY 12207

Portland Stove Foundry
57 Kennebec Street
Portland, ME 04104

The Stove Shop

A visit to a stove shop will open your eyes as to what is available. It may carry 50 to 100 makes and models in stock at one time. The majority of them will probably be for wood only or wood-coal.

Stove shops are found in most parts of the United States today, probably only a few miles from your home. Stoves are also sold by many department stores, hardware stores, lumber yards, garages, lawn mower shops and, yes, even grocery stores. In the long run you will be better off purchasing from the local, full-service dealer. He carries a good selection, is familiar with installation practices and local codes, and can supply accessories and spare parts. Many store owners will work with you to get a good installation that you are happy with, and will even take back a stove that doesn't do well.

Most manufacturers work through regional distributors. These distributors stock stoves, parts, and other supplies that the local dealer may not have space to store. They also provide technical assistance and training programs. The European stoves are brought in by importers. They serve the same functions as the distributors. In purchasing import stoves you will pay the duty fee and the added transportation cost.

Selecting a Stove

You can locate a dealer by referring to the telephone directory, weekly advertising newspaper, or contacting the manufacturers and distributors listed in the catalog section of this book.

Accessories

Some auxiliary equipment is necessary for the operation of your stove, other equipment is desirable. With each installation being a little different, the choice is up to you.

Tools for firetending. The coal hod or scuttle is convenient for moving and holding the coal. It has a long lip and can be used to pour coal directly on the fire. A small shovel and poker are also helpful for placing the coal and removing the ashes.

Heat extractor. This device attached to the stovepipe is used to remove extra heat from the flue gases before they reach the chimney. In installations where a very short pipe is used or a hot fire is maintained it serves a good purpose and it will pay to install one.

The simplest extractors are sheet metal fins attached or wrapped around the pipe. Blower-type units are also available that replace a sec-

Accessories for your coal stove

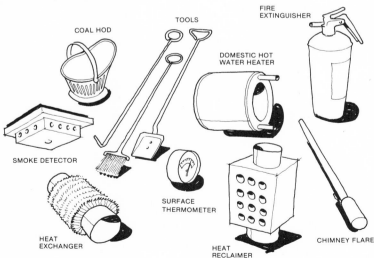

tion of pipe. These require electricity to operate but can be used to circulate heat within the room or duct it to an adjacent area.

Domestic water heaters. Add this unit to your stove or stovepipe and it can help reduce your electric or oil bill. The manufacturer's instructions must be followed exactly as these units can explode if not installed properly. At least one pressure-temperature safety valve is needed.

Smoke detectors. A must to allow you to sleep peacefully at night. They should be located above the basement stairs, near the ceiling in the hallway leading to the bedrooms, and in other locations where smoke may collect first. The cost of a detector is less than $10 and can be recovered in less than a year from the reduced premiums offered by many insurance companies.

Fire extinguisher. Locate several around the home, especially near the doorways to the kitchen and to the room housing the stove. They should contain at least 2½ pounds of chemicals and be rated to handle all types of fires. Getting to a fire in the first couple of minutes can often save extensive damage to your home.

False Advertising

With the stove industry rapidly expanding and stiff competition among manufacturers, some tend to exaggerate their design features or how well their stove performs. Reviewing a few of these may help you avoid some misconceptions about what size of stove you need or how it will operate. These examples have been taken from manufacturers' literature.

Example 1. "This stove will burn up to 100 hours on twenty-five pounds of anthracite," which means you can heat your home using approximately fifty pounds per week. Similar figures for an oil furnace would be to operate it not more than one minute each hour or use more than twelve gallons per month. I wonder how they survived the winter testing that stove.

Example 2. "Estimated heat output of Model X is 40,000–55,000 Btu/hr. Burn time—up to twenty-four hours. Average load—twenty pounds anthracite." A quick calculation will show you that their estimate is overstated by a factor of five. The actual output for an average load for the twenty-four-hour cycle using a 65 percent efficiency is only 7,500 Btu/hr. You won't keep warm with this one, either.

Example 3. "Heating Capacity 3,000–6,000 ft.^{3}." Be careful of the upper limit in this common statement. It assumes that your house is very tight, well insulated, and that the heat can circulate freely to all rooms.

Be careful of claims about the tests that a stove may have been given. These tests are only as good as the laboratory and the person conducting them. If the laboratory is not specified, ask for its name as it may be re-

---

CHECKLIST FOR SELECTING A STOVE

This list reviews the most important features that should be considered before selecting a stove.

1. Is the stove sized properly for the area to be heated? It may be too large if it will heat the area on the coldest day.
2. Is the style appropriate for the location where it will be placed?
3. A circulating type stove can be placed closer to a combustible wall and is safer around children.
4. Does the fireplace insert fit the size and style of fireplace?
5. Can the stove be used with the type of chimney available?
6. Is the stove listed or approved?
7. Will the stove burn coal efficiently? A wood stove will not.
8. Will the stove burn the coal that is available in your area?
9. Is airtight construction used for better draft control?
10. Was the stove built of cast iron or heavy steel plate?
11. Does the firebox have a firebrick or replaceable cast-iron liner?
12. Is a full shaker grate system used? Does the shaker work easily?
13. Can primary and secondary air supplies be controlled independently? The controls whether manual or thermostatic should be easy to adjust.
14. Is the fire door large and conveniently located?
15. Does the stove have a large ash pan or ash pit?
16. Does the stove have a durable finish?
17. Was quality workmanship used? Check for good castings, tight fitting doors, and continuous welds.
18. Is the stove warranteed for at least one year?
19. Are parts and service readily available?
20. Is a domestic hot water heater available?
21. Does the stove have a cooking surface?
22. Can the heat be distributed easily?

quired by the local building inspector. Statements such as "Meets UL Standards" or "Tested to be Safe by an Independent Laboratory" are not acceptable in most states.

Finally, check for a warranty. For some manufacturers who have little confidence in their stove it may only be for ninety days. Others give a fifteen-year warranty. Check to see what is included and excluded. Generally, items like the door gaskets, firebrick, and finish are excluded.

CHAPTER 3
A SAFE INSTALLATION

Safety should be of prime concern in a stove installation. You are familiar with the newspaper accounts of serious fires, injuries, and deaths that have been caused by an improperly installed stove.

In this chapter we will go through a step-by-step procedure for installing your stove so that you will feel at ease when you leave your home for the day or when you go to sleep at night.

Coal stoves do not have the safety features that are built into your oil or gas central heating system. The furnace or boiler has fused valves on the fuel supply line, high-temperature-limit switches, and pressure-temperature relief valves, all required by code and designed to handle emergency conditions. Central heating systems must also by code be installed and maintained by licensed service technicians.

Most state and national codes are vague in their requirements for solid fuel-burning devices. These codes were being developed when the use of stoves and heaters was declining. Now with the rebirth of these units the code writing agencies have not caught up yet. Jay Shelton points out in his book *Wood Heat Safety* (Garden Way Publishing) that further research is still needed on certain types of installations. Remember that you are installing the stove for the occasion when you forget and leave the draft open and it glows red hot.

The safety of a stove is not related just to its installation. Many fires are caused from improper use. Forgetting to turn down the drafts before you leave for work, leaving the evening newspaper too close when you retire for the night, or placing hot ashes in a cardboard box that is stored in the garage have all caused serious fires. The next chapter will deal with some of these aspects.

With care, coal stoves can be safe. Many homeowners have operated their stoves for years without any problems. Some leave their stoves operating all day while they work and find the house comfortably warm when they return. You will, with experience, develop a procedure that will allow you to do the same.

Who Will Install the Stove?

Should I install the stove myself? This question is often asked by the homeowner as he views the stove just delivered to his garage. Read through this chapter first before deciding, and consider the following:

1. Complexity of the installation. If it's just placing the stove on a stove mat and connecting the stovepipe to an existing thimble you probably can attempt it. If it involves extensive masonry work, a tall chimney installation, or other complex work you may be better off having a professional do it.

2. Skill level. Are you handy with tools? Have you had experience with projects like this before? Skill with the basic hand tools and several power tools is necessary to complete most installations.

3. Tools required. Do you have the necessary tools or can you get them to do the job? Make a list of these before you start. Often you can borrow them but you can save time by having the right tool when you need it.

4. How long will it take? A simple installation may involve just a weekend but the more complex job with masonry hearth, wall covering, and chimney could require all your spare time for several months.

If you feel competent enough and have the time, a stove installation can make a challenging and rewarding project.

Hiring a Professional Installer

Most stove shops have trained technicians to install stoves. One stove manufacturer has developed a UL-approved training program to certify these technicians. More manufacturers will probably adopt this practice as the advantages become apparent. Eventually most states will require certification or licensing just as they do for other heating system installers.

The professional installer is familiar with the local codes and permits required. He knows installation practices and where to locate the materials that are needed. You should inquire about the liability insurance he carries and what guarantee he gives on his work. Usually the guarantee is for six months or a year.

It's a good practice before hiring the installer to get two or three bids. These should be in writing and should list the materials and specify the work to be done. Once you have decided on the installer you should sign a written contract. This protects both you and the installer.

Stove Permit—Is It Needed?

The first step, once you decide to install a coal stove, is to see what codes apply. Some states, such as Massachusetts, Rhode Island, and Connecticut, have statewide building codes. In other states, city or town codes cover the installation of stoves. Check with your local building official, town hall, or city manager's office to see whether a permit is needed. In areas that don't have building codes, the local fire marshal or fire chief often makes inspections.

Where a permit is required a small fee, about $5, is charged for the inspection. This is probably the best insurance that money can buy. A signed inspection permit is valuable should you have a fire and need to collect from your insurance company.

There is another value to checking with the building inspector or fire marshal. They can often make helpful suggestions on the type of stove, its best location, and acceptable wall and floor protection, especially if you consult them before you purchase the stove.

Building codes are not interpreted or applied uniformly throughout a state or area. A good example of this is the homeowner in a nearby city who applied for a permit to install a coal stove in the basement and connect it to the oil furnace flue. He was told that this was not allowed. The next day the homeowner, in discussing alternatives with a different inspector from the same office, was told that the dual pipe installation was acceptable by code and that he would sign the permit. This problem is common in states with codes and makes it very difficult for dealers and installers.

The insurance company that issues your homeowner's policy may also require notification. For some it's just checking a box on the renewal form, others require an inspection either by their own agents or the local building official.

It's worthwhile inquiring about any rate increase. The way the insurance companies view the solid fuel stove movement is pretty much related to the number of losses they have suffered. A few with larger losses now require a rider be attached to your policy with an additional premium to be paid.

Hearth or Stoveboard

A decorative tile or brick hearth can be used to protect your floor and help to complement your coal stove. Combustible floor coverings must be protected from overheating and from flying sparks or hot ashes.

The National Fire Protection Association (NFPA) recommendations have been adopted by most building codes. The code bases it floor protection on the height of the legs on the stove. The taller the legs, the less protection you need. The code does not deal with coal stoves having the ash pit below the firebox. As this air space is insulation, most building inspectors measure the distance from the grate to the floor and use this as the leg height.

If you are installing a *listed* stove, follow the manufacturer's recommendations. The distances and materials suggested have been approved by the listing agency. If you are installing a nonlisted stove or don't have the manufacturer's directions, follow the NFPA recommendations shown in Table 7.

TABLE 7. FLOOR PROTECTION UNDER A STOVE

Air Space (Floor to Bottom of Stove)	Material
18 inches or more	24 gauge or thicker sheet metal. Available at sheet metal shop.
6 to 18 inches	¼-inch asbestos millboard covered with 24 gauge sheet metal. These stoveboards are carried by most stove shops.
Less than 6 inches	4-inch hollow masonry or tile with holes interconnected. 24 gauge sheet metal should be used above or below the blocks. Can be purchased at lumber yards, garden shops, or masonry supply companies.

If the stove is in the basement, garage, or other area with a concrete floor, no additional floor protection is needed.

Generally the materials suggested by NFPA are unattractive. Floor pads and hearths can be dressed up with materials that are noncombustible, continuous, and strong. The following are acceptable but should be placed *over* the basic material recommended by NFPA.

A Safe Installation **71**

1. Stone, brick, tile, or cast concrete blocks, mortared in place or set in sand.

2. Pea stone, trap rock, marble chips, etc. filled to a depth of at least two inches in a metal pan. Legs of stove must be supported firmly above the stone.

3. Copper or aluminum sheeting can be used to replace the sheet metal.

The size of the floor pad or hearth is also an important consideration. It should extend far enough from the sides and back to protect the floor from the radiant heat given off by the stove. The shorter the legs the greater this heat build-up can be. In the front or side, wherever the loading door and ash pit door are located, additional protection is needed. Sparks from a wood fire, especially soft woods such as spruce and hemlock, can shoot out quite a distance when the door is open. Also hot ashes spilled when they are being removed can burn holes in a wood or carpeted floor. NFPA requirements are six inches beyond the sides and back and eighteen inches in front of the loading door. Several state codes require twelve inches on the sides and back. Your installation will be safer using this dimension, especially with stoves having short legs or low grates.

If you are installing a fireplace insert, a small fireproof pad in front of the hearth may be all that is needed. Before placing a stove on the hearth

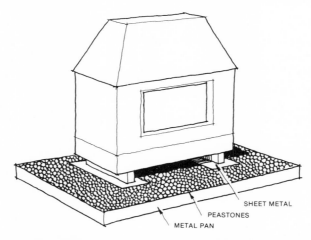

For an attractive stove base, a large metal pan is filled with pea stone. Note that stove legs are supported with pieces of metal that prevent them from sinking into the stones. For proper distances to be used in placement of stove on pan, see Table 7.

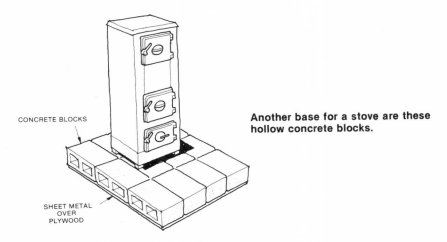

Another base for a stove are these hollow concrete blocks.

in front of the fireplace, check the hearth construction. The visible part may be noncombustible but may have been placed on the plywood subfloor or may have wood supporting members directly beneath it. A serious fire can occur from the wood overheating.

Protecting the Wall

A stove protruding out into a room is unattractive and creates problems of placing furniture and traffic flow, but minimum safe distances must be maintained so that combustibles do not overheat. In this section we will review these minimum clearances with and without added wall protection.

The materials that are potential fire hazards include wood, wallpaper, paint, paneling, curtains, and furniture. These can be ignited at temperatures as low as 200° F. when subjected to the constant heat of a stove. Other materials such as plaster, gypsum board, tile, and brick, although fire-resistant or noncombustible, are good conductors of heat. This means that the side touching the wall or studs may be just as warm as the face exposed to the heat. Any combustible touching one of these materials can char and eventually catch fire.

Two basic things must be considered when placing your stove: the type of stove, whether a radiant, circulating or cooking stove, and the type of wall protection used. If you are installing a *listed* stove follow the manufacturer's specifications even if they are less than those recommended by NFPA. The stove has been tested with the given distances and found to be safe. On the other hand, if you are installing an unlisted or used stove follow the recommendations given in Table 8.

Where wall protection is needed, the air space behind the sheet-metal protector is the key to the installation. It must be one inch minimum and more is desirable. Also an air space of one inch below and above the shield is required. This allows the air behind the guard to circulate by convection currents. The warmer the shield, the faster the circulation and the greater the cooling achieved.

TABLE 8. MINIMUM CLEARANCES FROM COMBUSTIBLES

Protection	RADIANT STOVE			CIRCULATING STOVE			COOK STOVE		
	Sides	Back	Front	Sides	Back	Front	Sides*	Back	Front
No protection	36	36	36	12	12	24	36/18	24	36
28 gauge sheet metal on ¼" asbestos millboard— no air space	18	18		6	6		18/9	12	
28 gauge sheet metal spaced out 1"	12	12		4	4		12/6	9	
28 gauge sheet metal on ⅛" asbestos millboard spaced out 1"	12	12		4	4		12/6	9	

Adapted from National Fire Protection Association Bulletin, 89M-1976

* Firing side/opposite side

The spacer material and location are also important. The spacers should be noncombustible, such as porcelain electric fence insulators, pieces of steel conduit or thin wall pipe or several small washers. The heat shield should be fastened to the wall with screws and spacers located no more than twenty-four inches apart. This will keep the shield from warping and reducing the one-inch clearance. Spacers should not be placed directly behind the stove. The heat shield could also be built free-standing or suspended from the ceiling by a chain or wire.

As with floor protection the materials recommended by NFPA are not attractive and alternatives can be acceptable. These include:

1. Replace the sheet metal with copper or aluminum sheet. It is possible to use strips of roof flashing and rivet or bolt them together.

74 HEATING WITH COAL

2. Cover one of the approved materials with a facing of tile, noncombustible artificial brick or a fire-resistant paint. Use a high-temperature adhesive.

3. Use commercial stove mats. Generally more than one are required to get the necessary size.

4. Build a ventilated brick or stove wall as a shield. It should be spaced at least one inch from the combustible wall. Use sheet-metal ties at least one per square foot to hold the shield to the wall. Allow air to circulate by leaving out half bricks in the top and bottom rows. Consideration should also be given to the added weight of the brick and whether additional bracing is needed to support the floor.

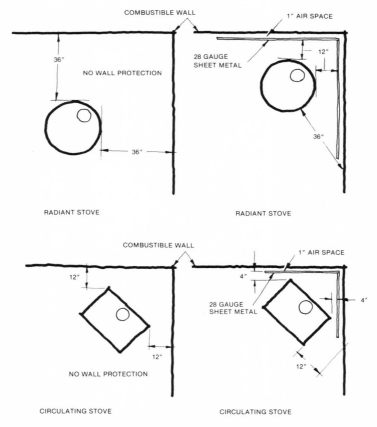

The placement of radiant and circulating stoves near walls, with and without wall protection.

A Safe Installation 75

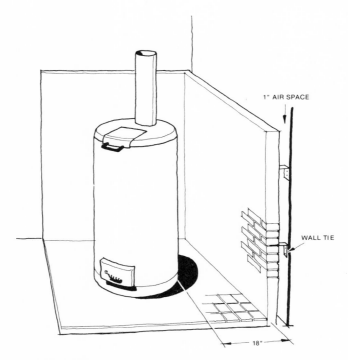

A ventilated brick wall is used as a shield in this corner installation. Note air space behind wall.

Size of Heat Shield

The heat shield should be large enough so that the bare wall is exposed only at a safe distance. These distances are listed on the "no protection" line in Table 8. For example, there should be no less than thirty-six inches from a radiant stove to any unprotected combustible wall.

Concrete is noncombustible but it is a good conductor. A heat shield is needed if you place a coal stove closer than twenty-four inches to a concrete wall that is faced on the other side with wood siding, as on the outside of the house or an insulated wall, such as between a garage and a basement.

Another question that comes up occasionally is the safe distance between a stove and a window or sliding glass door. With below-zero temperatures outside and 800° F. stove surface temperature the glass can be put to a lot of stress. The safe distance will depend on the size of the window and the expansion allowed for in its design, but a minimum of thirty-six inches should be used.

One other factor should be considered, that of placement of furniture, including magazine racks and wood boxes. These have been the cause of stove-related fires. Again three feet is the safe distance.

Stovepipe

The connection between the stove and chimney is often the weakest link in the stove installation. It is generally the simplest to install but one that causes many stove-related fires.

The NFPA standard for six- to nine-inch-diameter pipe is twenty-four gauge or heavier sheet metal. Most stove shops stock the lighter twenty-six and twenty-eight gauge because it is cheaper. Although these materials are probably just as safe when new, they deteriorate more rapidly from oxidation and the acids developed, and need to be replaced every year or two. The heavier materials are stronger and will last several years.

Stovepipe is available in several forms: galvanized, blue steel oxide and sheet metal painted with high-temperature paint or nickel plated. The galvanized pipe is not recommended for use in living areas as it can give off toxic zinc fumes if overheated, at least until the zinc burns off. The temperature at which zinc melts is about 750° F.

In planning your stovepipe installation try to make it easy and convenient to remove for cleaning and reassembly. Although creosote is not a problem unless you burn a lot of green or wet wood, you will have to remove the fly ash deposits from time to time. Pipe tees installed at strategic points can make cleaning and inspection much easier.

The following is a list of twelve installation pointers that will help you get a safe system:

1. Locate the stove so that the length of pipe between the stove and chimney is less than ten feet. This is more important with horizontal runs than vertical ones. Remember that the pipe is a good radiator of heat and that the surface area on a six-foot length of six-inch-diameter pipe equals the surface area of many stoves. This large surface can help to remove excess heat from the flue gases but in a coal stove operated at a low level it could reduce the draft to the point where the fire will die out.

2. Keep the number of elbows to a minimum. Each elbow tends to reduce the draft. Elbows are available in fixed and adjustable styles. With coal either one is acceptable. With wood the adjustable type may leak creosote.

3. Don't reduce the size of a stovepipe from the stove to the chimney as it will reduce the draft. If the stove collar takes a six-inch pipe, use six-inch or larger, never five-inch. If you have a European stove with metric-size collar, use either metric pipe or purchase an adaptor to the next size larger.

4. Stovepipe is not approved as a chimney and should not be used outdoors. Purchase a factory-built chimney for this purpose.

5. Stovepipe should be visible throughout its entire length. For this reason it is illegal to install it through a ceiling, floor, or in a closet or enclosed wall. If the pipe has to run through an interior wall a ventilated thimble or factory-built thimble should be used.

6. A bare metal stovepipe should not be placed closer than eighteen inches from combustible material such as a ceiling, wall, or curtain. If it must be closer use a heat shield. NFPA reduced clearance recommendations are shown in Table 9. The same alternate covering materials as discussed under "Protecting the Wall" can be used here to dress it up.

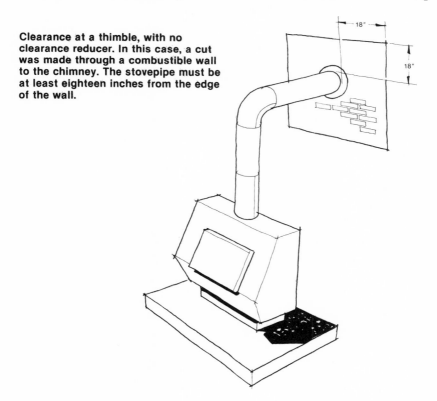

Clearance at a thimble, with no clearance reducer. In this case, a cut was made through a combustible wall to the chimney. The stovepipe must be at least eighteen inches from the edge of the wall.

TABLE 9. REDUCED CLEARANCES—STOVEPIPE TO COMBUSTIBLE

Material	Distance
None	18 inches
28 gauge sheet metal on 1/4-inch asbestos millboard; no air space	12 inches
28 gauge sheet metal spaced out 1 inch	9 inches
28 gauge sheet metal on 1/8-inch asbestos millboard spaced out 1 inch	9 inches

Adapted from NFPA standard 89M-1976, "Heat-Producing Appliance Clearances."

7. Stovepipe is sold in 24-inch lengths with a useful length of approximately 22½ inches. Often it comes unassembled to save packing space. To assemble, fit the corrugations together and compress the pipe to reduce its diameter. Once assembled, squeeze the ends of the joint with a pair of pliers. Pipe can be cut to length with tin snips or a hacksaw.

8. Connect sections together with three small sheet-metal screws in each joint. Holes should also be drilled in the stove collar for screws at that point. The screws keep the pipe rigid but still allow disassembly for cleaning. For coal stove installations it doesn't matter which way the crimped end faces. If you have a combination stove and plan to burn wood, face the crimped ends toward the stove to allow any creosote formed to drain back into the stove.

9. Stovepipe should never run downhill to the chimney or be installed lower than the stove collar; deadly carbon monoxide and other toxic fumes could escape into the room. It is better if it can be sloped up at ¼ inch or more per foot of length. If horizontal lengths of four feet or greater are used, the pipe should be supported with steel strapping or wire hung from the ceiling.

10. If a damper is not part of the stove, a separate damper should be placed in the section of stovepipe nearest the stove. Use a coal damper which has holes in it rather than a wood damper which is solid (although coal dampers are often sold for use with wood stoves). The holes allow some air to be drawn through to keep the fire going and the toxic fumes to escape even if the damper is closed.

11. A draft regulator similar to one used in an oil or gas furnace can be installed to equalize pressure if there is adequate draft in the chimney (0.05 inches water static pressure). This type damper should be located between the first two sections of pipe. Purchase one for either horizontal or vertical pipe installation.

12. Normally the joints between pipe sections are tight enough to operate the stove safely. In cases where the draft measured in the first section of stovepipe near the stove is marginal for operating the coal stove, applying furnace cement to the corrugated end before assembly can increase the draft slightly.

Chimneys

If we watch a hot air balloon gently lifting from the earth we observe that the only source of energy is the burner heating the air. As the air is heated it expands, becomes less dense, and the balloon is buoyed up by the denser atmosphere. A chimney works on basically the same principle. As the flue gases are heated they become less dense, rise in the chimney flue, and are replaced by air drawn in through the openings in the stove. This creates the draft or suction that keeps the fire burning.

Draft is measured by using a monometer, a clear plastic or glass "U" tube partially filled with water. One end is inserted into the stovepipe near the stove and the other left open to the atmosphere. The difference in the level of the water in the two sections of tube is the draft measured in inches. The draft in chimneys in most homes does not exceed 0.15 inch and is often near zero without a fire in the stove. Special gauges have been developed to measure these low readings. Furnace servicemen and stove dealers often use them when installing a stove or if there are smoke problems.

Several factors affect the operation of the chimney and thereby the operation of the stove. A basic understanding of these will help you to operate your stove better.

1. Increasing the height of a chimney increases the draft and the amount of flue gases that can be removed. This is why stoves in the basement generally burn better than those on an upper floor. The extra eight feet increases the draft by 0.03 inches at an average flue temperature of 300° F.

2. The hotter the flue gas the greater the draft. For most chimneys if the average temperature of the flue gases is kept 200° F. above the out-

door temperature there should be adequate draft to burn coal. To find the average flue gas temperature use a candy or meat thermometer and take readings in the last section of stovepipe before the chimney and at the top of the chimney. Add the two readings, then divide by two. Then subtract the outside air temperature.

Example: The temperature at the thimble is 300° F. and at the top of the chimney, 200° F., with a 40° F. outdoor temperature.

$$\text{Temperature Difference} = \frac{300 + 200}{2} - 40 = 210° \text{F.}$$

A stovepipe thermometer placed on the pipe can be used to find the temperature being maintained and, with a little experience, can be used to help control the fire.

3. The chimney, if it is too large for the stove, can affect the draft by creating too much cooling of the flue gases. For best results the chimney cross-section area should never be less than the area of the stovepipe collar and generally not more than 25 percent greater.

Problems often occur when trying to use a fireplace flue that is twelve inches or larger with a small stove having a six-inch diameter pipe connection. To correct this situation a stainless steel pipe six to eight inches in diameter is inserted the length of the chimney and connected to the stove. At the top of the chimney the pipe should stick at least four inches above a metal cap fabricated to cover the chimney.

4. Masonry chimneys, especially those on an outside wall, cool the flue gases more than an insulated factory-built chimney. This reduces the draft. It is possible when building a masonry chimney to add insulation such as vermiculite, perlite, or mineral wool between the flue liner and the brick to keep it warmer. There have also been cases where building an insulated enclosure around an outside chimney has improved the draft. This can be especially effective if the house is located on a windy spot and the chimney is on the north side. The enclosure can be built of framing lumber covered with siding to match the house. All wood should be kept a minimum of two inches from the chimney.

5. When a chimney is in an area of wind turbulance, the draft may be affected. Puff backs of smoke and fumes are the most common result.

6. Today's modern well-insulated homes are often so tight that not enough make-up air leaks in to keep a fireplace or larger coal stove operating properly. The large amounts of air needed by the fire have to be replaced. If not the fire will die down or puff backs will occur.

A Safe Installation   81

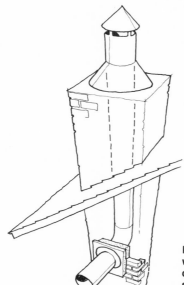

Installation of a stainless steel pipe within a chimney, used when a chimney flue is too large for the stove attached to it, or if chimney is unlined or in poor condition.

7. On damp, foggy days it seems difficult to get a good fire burning. This is because the air is heavy and requires more heating to create the same draft.

When considering the installation of a free-standing coal stove, one of the first things to check is the possibility of using a flue in the existing chimney. In older homes the chimney was often built with two or more flues. These generally served the furnace, a fireplace, and often a stove. The easiest way to determine what you have is to check the top of the chimney. Here all the flues will show. Smaller flues are used to vent the furnace and gas water heaters. Larger ones serve fireplaces. See if any of the flues are not being used.

If you find an unused one you may have saved yourself half the cost of the stove installation. The next step is to find the thimble or outlet. Often this is quite visible in the basement or a room on the main floor as a sheet metal cover near the ceiling. At other times. especially if remodeling has been done, the thimble may have been covered with Sheetrock or paneling, and you will have a harder time locating it. Try looking down from the top of the chimney using a strong light or mirror that reflects sunlight. You may be able to see the location and distance to the thimble where it enters the chimney. If not you may have to call a chimney sweep or mason to help you find it.

If you have an extra flue, make sure that it is safe to use. Often chimneys in older homes have deteriorated and need repair. In some homes you may find that a chimney fire cracked the liner or broke the mortar joints between the liner sections.

Checking the Chimney

The chimney needs to be inspected from top to bottom. If you don't mind this type of work including climbing on the roof you might save yourself $40 to $50 but it's often best to get a local chimney sweep who has the necessary equipment and know-how to do the job. Let's look at what needs to be checked.

- Is the chimney clean? You won't see much until you remove the soot, creosote, and birds' nests. A wire brush sized to fit the flue liner should be pulled up and down several times. The loosened soot and creosote should then be removed at the clean-out door at the base of the chimney.

- Chimney cap. The cap, if any, should be attached securely and be spaced at least three bricks' thickness above the top of the flue liner.

- The mortar that seals the bricks or stones to the flue liner at the top of the chimney should be sound and slope up toward the liner to help prevent down drafts.

- Flue liner. The tiles should be sound and not cracked. Mortar in the joints should not crumble when poked with a knife or awl. Give special attention to the mortar joints and the soundness of the bricks of chimneys without a liner.

- Chimneys with no flue liner or a cracked liner should be given a smoke test. Build a small, smoky fire, cover the top of the chimney and see whether smoke is escaping anywhere. A factory-built chimney or heavy-gauge stainless steel pipe can be used to line the chimney if the liner is unsafe. Be sure the joints are tight and the pipe is centered in the chimney.

- Clearance from wood beams. All beams and headers should be at least two inches from the chimney. In old chimneys where this clearance was not provided, sheet metal or aluminum foil slipped between the brick and the wood will give some protection.

- The clean-out door should be tight to prevent air from entering and reducing the draft.

A Safe Installation 83

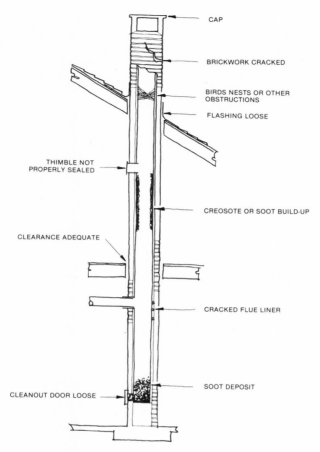

Check these spots when inspecting a chimney.

Dual Pipe Installations

Connecting a stove to a flue being used by another heating unit is allowed by some codes but it is not acceptable to most building inspectors and fire marshals. There are some good reasons for this. They include:

1. The flue may be too small to vent the gases from both the coal stove and the furnace, water heater, or other stove. When both units operate at the same time the gases may back up, forcing fumes and odors into the house. The efficiency of both units may be decreased because of a reduc-

tion in the draft. If one unit is shut off while the other operates, the chimney flue may be large enough, but other problems can occur.

2. Flue gases from one unit may enter the house through the second unit. Carbon monoxide from the coal stove and sulfur dioxide from the oil or gas furnace are toxic.

3. Deposits of soot and fly ash may build up in the bottom of the chimney, block one of the flue connections, and allow toxic gases to enter the room. If a dual pipe installation is put in, the primary heating device connections, usually the oil or gas furnace, should be placed above the secondary device, usually the coal stove.

4. Most oil furnaces leak a few drops of unburned fuel at times when valves do not close tightly. The air-gas mixture formed can be ignited by sparks or high temperatures from the coal stove and cause a puff back or small explosion.

It's best not to plan on using a flue already in use.

Using the Fireplace Flue

The simplest installation for many who don't have an extra chimney flue is to use the fireplace. There are several choices of heating units that can be connected to this flue. Before you purchase the stove, check with your local building official to see whether he will accept the fireplace flue as a chimney. Technically this flue does not meet all the requirements of a chimney and some officials will not accept it unless certain modifications are made. These may include bricking up the fireplace front and installing a separate thimble.

Also consider the ability of your hearth to take the weight of a stove. If at all in doubt, get professional advice.

In most cases the fireplace stove installation is a good one. It heats an area where much of the family activity takes place. It is generally in a room that is large and may have more than one door for good heat movement.

Several types of coal stoves will fit this location. Fireplace inserts that burn wood have been common for several years. Recently several manufacturers have designed units that burn coal. The insert described in the last chapter vents its flue gases through the open fireplace damper and up the flue.

How to Install a Fireplace Insert

Installation procedures include the following:

1. Check the chimney for soot, creosote, birds' nests, or other blockage. Clean it or have it cleaned by a chimney sweep if it is dirty.

2. Remove the fireplace damper plate. This can often be a difficult task as it may be corroded and rusted in place. The damper handle will have to be removed. It is usually bolted or pinned in place. If it doesn't drive out easily with a hammer and punch you may have to cut it with a hacksaw. The same procedure may be necessary to remove pins holding the damper plate. If all else fails and you can't get the plate off, you may have to wire it or block it open with a brick or piece of steel. Be sure that it cannot close; if it does, your house will fill smoke the next time you have a fire going.

3. Check the insert to be sure all parts are operating properly and then place it part way into the fireplace opening. See that its damper opening will be located behind the lintel, the piece of steel that supports the top of the fireplace opening.

4. Center the insert in the fireplace opening. Be sure that it does not stick too far into the opening. There should be at least a one-inch air space behind it to allow excess heat to escape up the chimney. Also there should be enough of the front of the insert showing so that the trim panels can be attached.

5. Mount the trim panels according to the manufacturer's instructions. These need to be sealed where they contact the insert. Generally a high-temperature silicone is used. It comes in a cartridge and should be labeled as No. 103 or 106. This can be applied with a caulking gun.

6. The trim panels must be sealed tightly where they contact the face of the fireplace. Fiberglass, mineral wool or another noncombustible material must be used. A strip one to two inches wide and about two inches thick should be glued around the back edge of the trim plates so that it can contact the face of the fireplace. Use the same cartridge of high-temperature silicone. In fireplaces having a rough face such as stone, additional insulation may have to be used.

7. Slide the insert back so that the trim plates form a tight seal. This helps to create a good draft in the stove. To check this seal place a flashlight up through the damper opening on the insert and look for light seep-

86 HEATING WITH COAL

ing through. It's best to do this at night with the lights turned off in the room. Add more insulation if needed.

8. Leave proper clearances around the insert. Follow the manufacturer's recommendations. One area that needs particular attention is above the insert. Generally twenty inches is required from the top of the insert to any wood mantle or wood trim. If you do not have the required distance some type of clearance reducer with an air space must be used.

Free-standing Stove in Fireplace

Often a free-standing stove connected to the fireplace flue is a better choice. A well-designed coal stove located on the hearth is more efficient than an insert, and the heat can be distributed more readily. With this type of installation a direct connection to the flue is necessary. Three types of installations are possible. Let's look at these in order of preference of installation.

Thimble above fireplace opening. This is the safest installation and the only one some building inspectors will accept. A thimble was installed above the opening when some fireplaces were built. If you have this, set the stove so that the stovepipe can connect to the thimble with a good

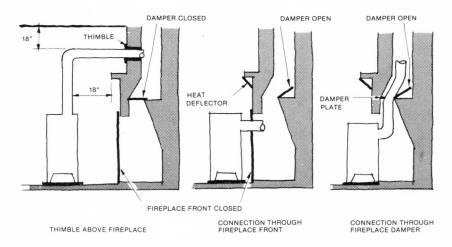

Three possible fireplace installations. The safest fireplace installation is at left, through a thimble and directly into a flue. In center, stovepipe is inserted through plate covering fireplace opening. This is not recommended. At right, stovepipe passed up through fireplace damper and into the flue. A method of preventing air from going up through the damper area is needed.

tight connection. Be sure that the pipe does not extend into the flue lining. If you don't have a thimble, you can get a mason to put one in. This involves removing a section of wall above the fireplace opening, drilling and chipping out the brick or stone, and drilling and cutting the flue liner. Once this is done the thimble is inserted and cemented in place. This is not a job for an amateur. Care must be taken so that the flue liner is not cracked or chipped. A good seal must be made at the flue liner–thimble interface so that creosote that forms when wood is burned will not leak between the flue liner and the brick.

A combustible wall *must* be protected if a thimble goes through it. If you use brick around the thimble, it must extend eight inches out on all sides. This means that for an eight-inch thimble, you must provide a hole twenty-four inches square. If an air space is left, the distance between the thimble and the combustible wall must be at least eighteen inches; an eight-inch thimble will require a hole forty-four inches square.

This is the most important area in the stovepipe connection. Many fires get started when the thimble area overheats and ignites the combustible wall that covers the chimney. Check the distances carefully.

Once the thimble is in, see that the flue is clean. Then seal the fireplace

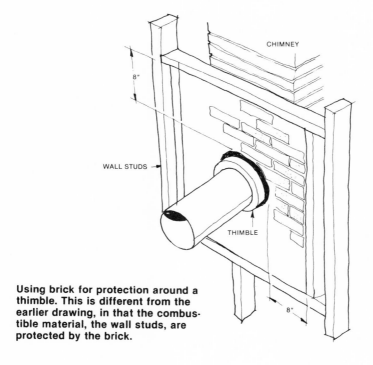

Using brick for protection around a thimble. This is different from the earlier drawing, in that the combustible material, the wall studs, are protected by the brick.

damper. Often dampers are warped or cracked from overheating and do not close tightly. If yours does not, cut a piece of sheet metal or asbestos board and attach it over the opening, using sheet-metal screws, angle brackets, or clamps. Small cracks can be filled with furnace cement or fiberglass insulation.

Locate the stove so that it is a safe distance from combustibles according to the manufacturer's instructions. If the hearth does not extend out far enough use a stoveboard. Also make sure that the stovepipe clears any wood mantle by eighteen inches. Otherwise a clearance reducer, such as a heat deflector, will be needed.

Connection through fireplace damper. In this installation the stove is set in front of or into the fireplace opening and the stovepipe connects through the damper. The stovepipe must extend up through the damper plate into the first section of the flue pipe, above the fireplace shelf. This improves the draft.

Steps to be taken:

1. Remove the damper plate as described earlier or block it open.

2. See that the chimney is clean and in good repair.

3. Fit a piece of sheet metal, twenty-four gauge or heavier, to cover the damper opening. This must have a hole in it to receive the stovepipe. Several methods can be used to hold this plate in place, depending on the fireplace construction. Some use screws, expansion bolts, or brackets. Others use a boxlike piece of sheet metal. To make one of these, take the measurements of the opening just below the damper. Add six inches to these dimensions and cut the sheet metal to that size. Then cut out three-inch squares for the corners, and bend the three-inch flanges up less than 90°. Cut an opening for the pipe. Insert the metal sheet into the opening, flanges down so they will hold the piece in position. For an airtight installation, caulk around the flanges with furnace cement. Do the same around the stovepipe when it is in position.

Some hardware stores and welding suppliers are making these units, usually with a small section of pipe slightly larger than your stovepipe welded to the sheet metal. That short section of stovepipe has holes for sheet metal screws which will secure the pipe to the unit.

Not all damper openings are large enough to take a six-inch stovepipe. If the pipe doesn't fit or if you have a larger pipe, you will have to compress it into an oval, and also cut an oval hole in the plate.

4. Connect a couple of stovepipe sections to reach from the stove through the damper plate to the flue. Use three sheet-metal screws per

section to hold the sections together. Add sections one at a time until the far end is at least six inches into the flue lining. Use adjustable elbows to make any necessary angles.

5. Fill cracks around the damper plate with furnace cement or fiberglass to get a tight seal.

6. Move stove into position, being sure to maintain the proper clearances as given in the instruction manual or the section on clearances in this chapter.

7. Connect the stovepipe to the stove and fasten into place. The unit should be easy to disassemble for cleaning.

Some stove dealers recommend that fiberglass insulation be used around the pipe at the damper. This should not be done for three reasons:

1. Most fiberglass will melt at the temperatures reached in stove pipe.

2. Fiberglass does not form a tight seal and the draft will be reduced.

3. The fiberglass may fall out, especially if it becomes wet from a heavy rain.

Stovepipe inserted through plate covering fireplace opening. This is the least desirable of the fireplace installations. The flue gases being exhausted into the large opening are cooled and the draft may be reduced too much to get good burning. If you want the appearance of the closed fireplace, follow the recommendations of the previous installation with the stovepipe inserted up through the damper opening and then before assembly add the cover plate. This plate should be six inches wider and three inches taller than the opening. It should be made to attach easily to the stove or brick for easy removal. A tight seal such as fiberglass glued to the metal with high-temperature silicone adhesive should be used.

The sheet metal can be painted with high-temperature paint or it can be covered with Z-brick, tile or other noncombustible facing material.

A New Chimney

If you have to install a separate chimney you have several choices—masonry, faced either with brick or stone; cement block; and factory built. A masonry chimney within the house acts as a heat sink absorbing the heat from the fire when it is hot and then giving it back when the fire cools down. External masonry and factory-built chimneys don't do this.

Materials for a factory-built chimney usually cost twice as much as for a cement block or brick chimney, but if you are handy with tools you can install the pre-fab yourself.

Masonry Chimney

Building masonry chimneys is better left to a qualified mason. You can see why if you look at the steps that must be taken to build one.

- Dig a hole for the foundation. It must extend below the frost line outside or have a good footing if located within the house. The foundation must be at least a foot wider and longer than the chimney size.

- An ash pit door must be inserted into the chimney and connected to the flue near the base.

- The flue liner must be surrounded by an air space and a layer of brick or stone. A two-inch gap must be left between the chimney and combustibles. The external chimney must be tied to the house.

- A thimble has to be cut into the flue liner and extended through the wall. All combustibles must be kept a safe distance away.

- A special mortar is used between the flue liner sections.

- The top of the chimney should be built to shed water and prevent down drafts.

- Flashing has to be installed where the chimney goes through the roof or overhang.

Factory-Built Chimney

A factory-built chimney is a good choice for many coal stove installations. These are available as double-wall insulated pipes or triple-wall air-cooled pipes and work on the principle that air spaces or insulation between the walls will reduce the surface temperature to a safe level. Only pipe that has been approved by Underwriters Laboratory or other testing services should be used. These pipes are tested at 1700° F. for ten minutes and for longer periods at lower temperatures.

Only "All Fuel" or "Class A" chimneys should be used for coal or wood stove installations. These have a stainless steel inner liner to give long life under corrosive conditions. A similar-looking "Gas" factory-built chimney is available but must not be used with solid fuels.

Factory-built chimneys are tested using the installation instructions provided by the manufacturer. These instructions should be followed exactly. Most pre-fabs require a two-inch clearance to combustible material.

Double-wall pipe uses a high-temperature insulation between the walls, generally silica and fiberglass. Triple-wall pipes have only air, either a dead air space with caps at both ends of each section or a ventilated air space with the cool air moving down through the outside space and up through the inner space. With the thermosyphon type the hotter the flue gases become the greater the air movement and the greater the cooling. Although some states question the use of the thermosyphon type for a wood stove because of the possibility of greater cooling and more creosote build-up, all types will work well with coal.

The factory-built chimney can be installed with hand tools. A few pieces of 2 × 4, some nails, and roofing cement will be needed. The pipe comes in eighteen- and thirty-inch lengths. Other items you need, such as elbows, tees, caps, and support brackets, are available. Let's look at the two most common types of installation.

Installing Through the Ceiling

In this installation the stove is generally located centrally in the house. Single-wall stovepipe is run from the stove to within eighteen inches of combustibles, then pre-fab runs through a ceiling support bracket and then through the ceiling, the upstairs room, if any, the attic, and the roof.

Locate the stove to have safe clearances from walls and other combustibles. Keep the location near the center of the house so that the chimney will exit near the roof ridge. This will eliminate some bracing and the tall exposed chimney that would be required if it were closer to the eave. Also check in the upper floor and attic to see that there are no obstructions in the general area where the chimney will be located. Fit it between the ceiling joists and roof rafters. If you cut one of these you may weaken the house, and will have to do some extra bracing. If necessary the chimney pipe can be tilted an inch or two, or you can install a less desirable offset in the chimney. Use fifteen- or thirty-degree elbows for this.

Caution: This type of installation cannot be put through a ceiling with radiant electric or water heat. Also be careful in cutting areas that may have electric wiring in them. It's best to remove the fuses or turn off the circuit breakers that power lights or wall plugs in the room you are working in.

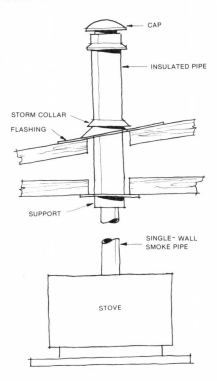

Prefabricated chimney installed through roof

Before cutting, locate the position of the hole in the ceiling and roof. Pilot holes drilled through the ceiling can be used to locate the position on the upper floor.

A keyhole saw or sabre saw can be used to cut away ceiling and floor material. If you have to cut a ceiling joist or roof rafter support the area with jacks or 4 × 4 posts before cutting. Use plenty of nails in the headers.

All openings must be framed. Use 2 × 4 or 2 × 6 lumber for this and toenail them into the framing. This supports the edges of the ceiling or floor material.

A ceiling support package is used to hold the factory-built chimney. It is held in place with nails or screws. Maintain the proper distance to the wood around it.

After removing the roofing and roof sheathing, add the framing. Next attach the roof support bracket and the flashing that prevents the rain from getting into the attic. Use roofing tar to get a watertight seal. Apply tar to loose ends of roofing and to exposed nailheads.

A Safe Installation 93

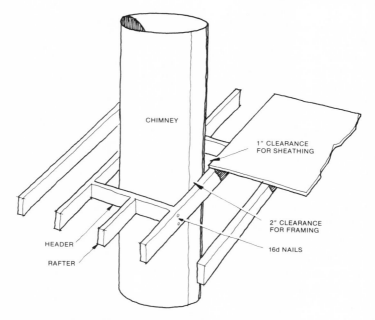

Roof framing for prefabricated chimney

Put the section of pipe which will pass through the roof in place, slip the storm collar over the pipe, and then add sections of pipe until you are at least two feet above the roof and two feet above the ridge or any part of the roof within ten feet. If the chimney extends more than six feet above the roof use a chimney bracket or guy wires for support. Finally add a cap to prevent down drafts and rain from affecting the draft. Fake masonry housings are available to cover the exposed part of the factory-built chimney. These do not affect the operation and are purely decorative.

Installing Through the Wall

This type of installation is generally simpler than the ceiling type because only one hole has to be made. You will be working on the outside of the house where there is more room. You will have to work from a ladder and care should be exercised in its use. A second person should help in assembling the pipe sections and holding the ladder.

In this installation the stove is located near an exterior wall where the chimney can be placed without blocking upstairs windows or interfering with utility lines. If possible, locate the chimney where it will not have to extend too far above the roof and become unsightly.

Place the stove in a position to maintain safe clearances from combustibles. If you need to get closer than thirty-six inches to combustible walls use an approved clearance reducer.

Use a tee rather than an elbow at the base of the chimney outside the house to make cleaning easier. Try to locate the pipe so that you will not have to cut a wall stud. You can locate these with a stud finder, or by tapping lightly on the wall with a hammer. A hollow sound indicates the space between the studs.

If a stud must be cut, place a 2 × 4 header above and below it at the correct spacing for the wall tee. Toenail and end nail the headers in place. Drill pilot holes through the outside wall so that you can cut away the exterior siding and sheathing. Be careful of nails and electric wiring.

Attach the wall spacer and wall bracket. Make sure that the bracket is securely fastened as it will support the weight of the chimney. Fit the tee into place and then assemble chimney sections and exterior wall supports as needed. If you have a large overhang you can either cut through it and use flashing and a collar, or you can use fifteen- or thirty-degree elbows to offset the chimney around the overhang. Use the approved wall standoffs and keep the recommended clearances from combustibles.

Using a tee at the bottom allows the chimney to be inspected and cleaned

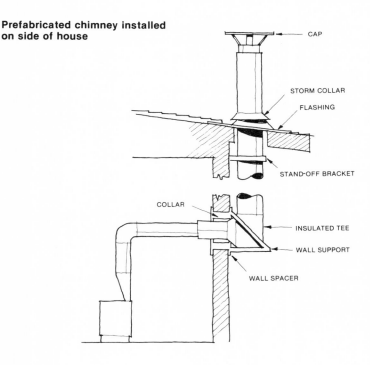

Prefabricated chimney installed on side of house

as necessary. The plug at the bottom of the tee should be tight so that a good draft is maintained. A cap should be used at the top of the chimney. As with other installations the chimney top should be at least two feet above the roof and two feet above anything within ten feet horizontal distance.

To complete the installation, connect the stovepipe from the stove to the chimney. Be sure that the connection to the chimney is tight and that three sheet-metal screws are used at each joint.

If you are in an area where lightning protection is needed have the chimney grounded according to the requirements of the local code. Grounding wires should be connected to a water system having metal pipe or to copper-coated rods driven into the ground.

Heat Distribution

In several sections we have discussed locating the stove so that the heat is distributed evenly throughout the area. In one or two rooms natural circulation will do the job. In larger areas stoves with circulating blowers or small fans can be used. When the stove is in the basement distribution becomes more difficult. A few manufacturers make stoves with connections for external ducts. A blower on the stove moves the heat.

Some homeowners install a sheet-metal hood over the basement stove to capture the heat and direct it through ducts to the upstairs rooms. These ducts, usually standard heating system ducts or stovepipe, should be insulated with one to two inches of fiberglass to retain the heat. They should also be kept at least one inch away from combustibles. A cold-air return is needed from remote sections of the house to get good air movement.

Make-up Air

Large volumes of air are needed to burn coal. Some of this air is needed for combustion. Most of it, especially in fireplaces and fireplace inserts operated with the doors open, is extra air that removes heat from the fire and the house. To replace it, air leaks in through cracks around doors, windows, and vents and can cause cold drafts on your feet.

The use of a separate duct to bring in outside air and direct it to the area of the stove can eliminate these drafts. It can also help alleviate problems in our modern well-insulated homes. Sometimes the operation of a fireplace or large stove can cause venting problems in oil or gas furnaces

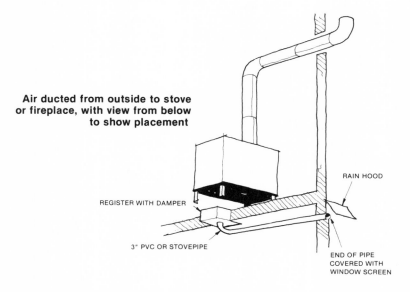

Air ducted from outside to stove or fireplace, with view from below to show placement

or water heaters. The make-up air for these units could be drawn from the chimney, causing smoke and toxic fumes to be brought into the house. In very tight houses operation of kitchen or bathroom exhaust fans can cause the same effect.

The claim that use of ducted air can increase energy efficiency by not using heated air to operate the stove has not been proven. The efficiency of the coal stove usually decreases as the fire runs cooler and heat transfer is reduced.

If you need to get extra air into the house you should use four-inch stovepipe or PVC drainpipe. It should be located below the level of the stove. Often it is best to run it through the basement area. The pipe should be insulated so that moisture doesn't condense on the cold pipes. Place a piece of window screen over the outside end and have a louver

---

GAS SNIFFER

There is a safety device, which detects the presence of dangerous concentrations of carbon monoxide, propane, gasoline and other de-oxidizing atmospheres, that prospective coal stove owners may wish to consider. Revco Products manufactures one of these devices, the Gas Sniffer. To locate the distributor in your area, write to Revco Products, 1530 Edinger #6, Santa Ana, CA 92705.

fabricated for the inside end, so that it is not piping cold air into the house when that air is not needed. This end should be located near the stove but not where hot ashes can fall into it. A make-up air duct is often built into new fireplaces.

Mobile Home Stoves

Several precautions must be taken when installing a coal stove in a mobile home. These homes are built using different materials and methods than are common with conventional homes. The strength of the floor and greater flammability of the materials may affect the stove installation.

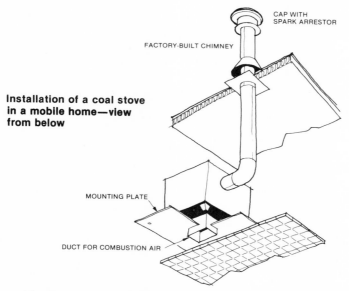

Installation of a coal stove in a mobile home—view from below

Check with the local building inspector or fire marshal for regulations that apply to these installations. In most cases the Department of Housing and Urban Development (HUD) requirements supersede other regulations. These include the following:

1. The weight of the stove must be distributed so that it does not exceed forty pounds per square foot. This can be done with a large sheet of ¾-inch or one-inch plywood under the floor pad.

2. It should be possible to bolt the stove to the floor when the mobile home is moved.

3. The factory-built chimney must be attached to the stove directly. NO stovepipe. A spark arrestor must be installed.

4. Air for combustion must be ducted to the stove from outside.

Some manufacturers make stoves that meet these additional requirements. Purchasing one of these units will greatly simplify the installation and is worth looking into. Because mobile homes have relatively small volumes, select a smaller stove; otherwise you may have to keep the windows open to keep the temperature at a comfortable level.

Zero-Clearance Fireplace Inserts

Zero-clearance fireplaces are designed to be installed directly against combustible material. This is done with two or more air spaces and insulated barriers around the firebox. Sometimes a separate housing is designed that can be fitted with several stoves. Some units operate with a blower to distribute the heat and cool the firebox.

Zero-clearance units are designed for areas where space is limited. They are often used in modular pre-fab homes, condominiums, and in the corners of small rooms where one does not want a stove to stick very far out into the living space. As with other fireplace units, unless they are designed specifically to burn coal the performance from this type unit will be poorer than from a well-designed coal stove. Most inserts use a cast-iron grate set into the firebox.

Precautions

In installing a zero-clearance fireplace there is no tolerance for error. The manufacturer's instructions must be followed without deviation. These instructions were used when the unit was tested and approved by UL or other approved testing agency. The installation precautions include the following:

1. Comply with local building codes.

2. Use only the chimney recommended by the manufacturer. Maintain the approved clearance for the chimney. It is generally two inches minimum. Stovepipe is not allowed.

3. Don't damage the unit while it is being installed.

4. The framing around the unit is best constructed after it is set in place and the chimney installed.

5. Any hot air ducts must be wrapped in insulation. Use the material provided or the recommended material.

6. Do not block off any hot or cold air duct.

7. If a blower is included as part of the unit, have a licensed electrician make the connection.

8. Provide the required hearth protection.

CHECKLIST FOR A SAFE COAL STOVE INSTALLATION

Use this checklist before starting the first fire in your stove. For more detailed discussion of the installation procedures refer to the text.

1. The stove is listed or approved for use in your state.

2. The stove does not have broken parts or large cracks that make it unsafe to operate.

3. The stove is placed on a noncombustible floor or an approved floor protection material. This material extends at least six inches from the sides and back of the stove and eighteen inches from the front where the coal is loaded and ashes removed.

4. Radiant stoves are placed at least thirty-six inches away from combustibles. Circulation type- twelve inches. If not, an approved clearance reduction material and air space are used.

5. Curtains, furniture, and other combustible material are at least thirty-six inches from the stove.

6. Stovepipe of twenty-two or twenty-four gauge metal is used. Three sheet metal screws are used in each joint.

7. The diameter of the stovepipe is not reduced between the stove and chimney flue.

8. A *coal* damper is installed in the stovepipe near the stove.

9. Length of the stovepipe is less than ten feet. No more than two elbows are used.

10. There are at least eighteen inches between the stovepipe and combustible material or an approved clearance reduction material is used.

11. The stovepipe slopes upward toward the chimney and enters the chimney higher than the top of the stove exhaust collar.

12. In masonry chimneys the stovepipe enters the thimble horizontally and does not extend into the flue. A tight joint exists between the stovepipe and thimble.

(continued)

Checklist for safe installation (cont'd)

13. A UL-approved all-fuel factory-built chimney is used where a masonry chimney is not available or practical. It is installed according to the manufacturer's recommendations.

14. The stovepipe does not pass through a floor, closet, or concealed space. The stovepipe does not enter the chimney in the attic, closet, or enclosed area.

15. A double-walled ventilated metal thimble is used where the stovepipe goes through an interior wall.

16. The chimney is in good repair.

17. The chimney flue and stovepipe are clean.

18. In a fireplace insert installation, the seal between the trim plate and the fireplace face is tight.

19. The stovepipe for a stove installed into a fireplace extends into the first section of flue lining above the shelf area. The damper connection is tight.

20. A metal container with a tight lid is available for ash disposal.

21. The building official or fire marshal has approved the installation.

22. The company insuring the building has been notified of the installation.

23. A smoke detector is installed near the ceiling in an area adjacent to the stove.

24. A fire extinguisher (at least 2½ pound ABC type) is near the entrance to the room with the stove.

CHAPTER 4
FIRING THE STOVE

Now that you have made a final check you are ready to put the stove to work. If you have had experience with burning wood and are switching to coal you will find that the procedure is similar except that the coal reacts more slowly. If the last time you lit a fire was when you were a Scout the learning process may take longer and require more patience. The burning of coal is as much an art as it is a science and each stove installation will have its own peculiarities. These you will learn only with experience.

If you know some of the basic principles, it will be easier to operate your stove and make the proper adjustments. Coal, having much less volatile matter than wood, requires more air from the primary supply below the grate. This air supply controls the rate of burn and the amount of heat that you get from the stove. The larger the opening the faster the coal will burn.

It's also important to maintain an adequate temperature in the firebox. With a wood fire a low firebox temperature indicates a smoldering fire with a lot of smoke and the formation of creosote. With lignite and bituminous coal the fire tends to be smoky. If you are using anthracite, which is mostly carbon, the fire may not generate enough heat to keep itself going.

To get high efficiency from the fuel you burn there must be good mixing of the air and the combustible material. This is partly controlled by stove design but can be affected by the operation of the dampers. More secondary air is needed in the lower-rank coals. Also higher temperatures and greater mixing is obtained by controlling the damper in the stovepipe to the point where the gases are burned before they enter the chimney.

When coal starts to burn, the volatile matter is driven off first. This may take only a couple of minutes with anthracite or ten to fifteen minutes with the high-volatile bituminous and lignite. During this phase of the burn cycle additional heat can be obtained by allowing more air above the fire where these volatiles form. Once they burn the secondary air supply can be cut back.

What you have left is mostly carbon or coke. This requires a higher temperature to burn, generally 1500° to 2000° F., and a greater air supply from the bottom.

What Size and Type of Coal?

Use the size and type that the manufacturer lists in the instruction manual. Generally the coal recommended is the type available in the area where the stove was built.

If you have a second-hand stove or one that was made in another part of the country or world, what should you burn? In general you can with some experience burn almost any type of coal in any coal stove. There are a few features that differ and should be noted. The lower-ranked coals—lignite and subbituminous—have less heat value per pound and stoves designed for these fuels often have larger fireboxes than stoves for anthracite that give the same heat output. Provisions for supplying more secondary air are needed for the low-rank fuels because of their greater amount of volatile matter. Stoves with thermostatic control should also be checked for this point. Those having a primary-to-secondary air supply ratio closer to 1:1 were probably designed for lignite or subbituminous, or as a combination wood/coal stove.

The size of coal to use depends on the diameter or cross-section of the firebox, the depth to which the coal is fired, and the size of the grate openings. Larger sizes of coal can be used with larger stoves having a greater surface area and heat radiation.

The resistance to the primary air flowing up through the fire bed changes with the size of coal. Larger pieces don't pack as tightly together and so allow an easier air path. If you are having a problem of poor burning caused by insufficient draft, try using the next larger size coal.

The grate openings must be small enough to support the coal without allowing much to fall through to the ash pit. Generally the largest grate opening dimension should be at least ¼ inch smaller than the minimum size of the coal.

The following recommendations were adopted many years ago by the anthracite industry.

Stove coal is suitable where the firebox is sixteen inches in diameter or greater and twelve inches or greater in depth.[1] It is generally used in furnaces and only the largest of heaters.

---

1. Depth refers to the distance from the top of the grate to the bottom of the fire door.

Nut coal is the best choice for most stoves. It can be used where the firebox is up to twenty inches in diameter and ten to sixteen inches deep. It is also ideal for most kitchen ranges and fireplace grates.

Pea coal is used in the smaller stoves, kitchen ranges, and in other stoves having adequate draft. It is used in mild weather and in banking for the night.

Buckwheat is the smallest size that can be burned with natural draft. You must have an exceptionally strong draft to get satisfactory results.

Recommendations for the size of *coke* to use are as follows:

No. 2: Nut —small stoves, kitchen ranges, hot water heaters
No. 1: Nut —larger stoves, heaters
Egg Size   —furnaces and boilers

Bituminous coal and lignite are graded into many more sizes than anthracite though generally only two or three sizes are available at the local coal yard. The coal dealer can help you decide which of those sizes he carries will work best in your stove.

Operating the Stove

For convenience, this section will be divided into the various types of fuels because the operating procedures vary with the type of fuel used.

All coal fires should be started with wood. This is needed to get the fire hot enough to ignite the coal. Charcoal, although easy to use and readily available, is expensive and may give off toxic fumes; it is not recommended. The wood should be dry. Softwoods make good kindling because of the resin they contain and the fact that they split easily. Hardwoods are better on top of the kindling to give a longer-lasting fire until the coal gets started. If you operate your stove so that the fire only goes out a few times during the season a car trunk full of kindling will be all you need.

If you have a new stove you should break it in slowly for about a week. Have a small wood fire the first few times and then a small coal fire to allow the castings to cure. You may see some condensation form on the stove from the moisture in the firebrick. Also you will smell the fumes from the oils in the paint. These are normal occurrences and you shouldn't worry about what is happening. It is best if you can light a small fire in the stove outside before you install it to burn off the oils. Otherwise they may smell up the house. Small fires give you the chance to get a feel for the stove and how it operates. Experiment with using the dampers, adding the wood or coal, and cleaning the ashes. Also watch for any smoke leaks from the stove or stovepipe. If they occur you can usually seal them with furnace cement.

Burning Anthracite

Anthracite is the most common coal used in the home. Its long, even burn time, high heat output, and cleanliness make it a good choice for the home. Because it is used in industry, it is readily available in most parts of the United States. It is generally more expensive than other types.

Starting a Fire

Before starting the fire, open the smoke pipe, fire door, and ash pit dampers. Place some crinkled newspaper and finely split kindling on the grate. Crisscross the kindling to allow air to get through. Use dry softwood. A few scraps of pine boards will get the fire started quickly. Light the paper. Add larger wood in a few minutes after the kindling is burning brightly.

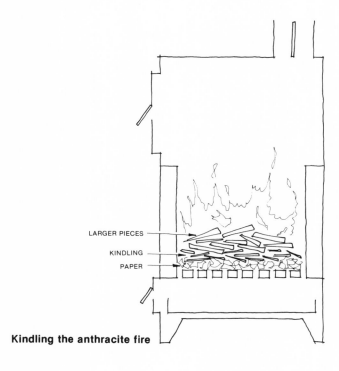

Kindling the anthracite fire

Caution: Never sprinkle gasoline, kerosene, or other explosive liquid over the kindling. This could cause serious injury.

Place the larger pieces on the fire so that they are slightly separated and form a level bed for the coal. It will take ten to twenty minutes before they are thoroughly ignited and ready for the coal. Adding the coal too soon could cut the air supply and smother the fire.

Add a thin layer of coal, preferably smaller chunks, to the wood fire, being careful not to disturb it too much or cut off the draft. Add a second, heavier layer when the coal is ignited and burning well. In stoves with a shallow firebox this may be all you will need to add until recharging. With deeper fireboxes, a third layer may be needed to bring the coal up to the bottom of the fire door. In both cases you should leave a red spot of glowing coals visible after firing to be sure you haven't smothered the fire and to help ignite the gases given off by the new charge. A deep charge will give a more even heat and a longer fire. It may take one to two hours before the whole bed is fully ignited.

When the fire is well established and the room is becoming warm, partially close the dampers. The secondary air supply, usually on the fire door, can be nearly closed. Leave the ash pit damper partially open,

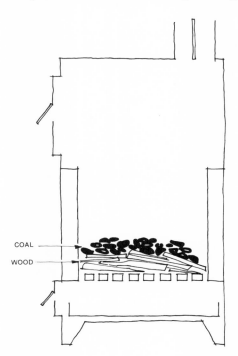

Adding coal

otherwise the fire will go out. Adjust the stovepipe damper to reduce the draft on the fire. With most types of anthracite you will see short blue flames above the coal except when the fire is started or a new charge is added. If there is no flame the fire needs more air from the bottom or it is near the end of its burn cycle and needs to be recharged.

Refueling

When the coal has burned down to half its original depth it's time to add coal. Open the dampers and allow the fire to pick up a little and also burn off the gases. Open the fire door and pull the glowing coal to the front of the firebox. Try not to disturb the fire too much. Next add a fresh charge at the back, being careful not to seal off the top. Close the fire door but leave the fire door damper open for a few minutes until the volatile gases have burned off. It is not necessary to shake down the ashes each time you refuel most stoves, but only experience can tell you this.

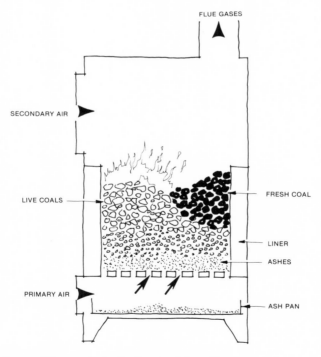

When refueling, pull live coals to the front, add coal in rear.

Shaking the Ashes

Be gentle when you shake the ashes. A few short shakes are better than a large movement of the grate. The objective is to remove a small amount of the ashes without disturbing the fire. The fire should just be settled down about a half an inch or an inch in the firebox until the first live coals start to fall. Excessive shaking wastes fuel by allowing unburned coal to drop into the ash pit. It can also expose the grate to very high temperatures, which can warp or burn it out. The fire may go out if you shake it too much.

With stoves that do not have a full bottom shaking grate you may have to get in about every other day with a poker and push down the ashes that build up in the corners. This has been a problem with many of the newer stoves where the manufacturers have skimped a little and have only placed a shaker in the center of the grate. These stoves will operate fine if you realize what the problem is and take the corrective measure.

A few imported stoves have no shaker in the grate. You have to use a thin poker or *fiddle stick* made from coat hanger wire or steel strapping 1/8 inch thick by one inch wide. This is placed on top of the grate and moved back and forth with a short choppy motion to knock the ashes through. This type of grate system will work but it is not as convenient as a shaker grate.

You should also be careful in shaking the ashes so that you don't form clinkers. These form when the very hot burning coal comes in contact with the ash layer. This occurs when you shake the fire too much or poke a fire. Some coals, especially those high in iron, form more clinkers. Because clinkers will not burn and will block the grate when formed in large pieces, remove them as necessary before refueling.

Banking the fire. For overnight operation or if you are away all day, you will want to bank the fire. To do so, heap the coal up along the sides and back of the fuel box so that the fire gradually burns it over a longer period of time. You also reduce the intensity of the fire without letting it go out. Follow the same procedure as for refueling. If possible, avoid shaking, as a heavier layer of ash will help reduce the intensity of the fire during this time. In stoves having poor draft control you can get longer burn cycles by using a layer of smaller pieces of coal, such as the fines that collect near the bottom of the pile. This will reduce the combustion rate by impeding the flow of air.

After loading, let the fire establish itself for about a half hour and then close the dampers to the point where the house does not become too cold.

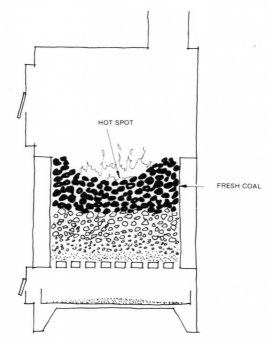

When banking the fire, heap the coal up along the sides and back of the fuel box.

As you can see, it's important that the banking be done early enough before you retire or leave so that you can make adjustments after the fire is well established.

Reviving the fire that's almost out. Occasionally you may find that the fire is almost out before you remember to refuel it. You may first notice this as the house cools. The first thing to do is open the ash door and stovepipe dampers and close the fire door damper to get a good draft through the grate. Then place a thin layer of dry coal from the top of the pile over the entire top of the fire. *Do not poke or shake the fire at this time.* After the fresh coal has become well ignited, shake the grates and refuel.

After the fire goes out. This will happen from time to time even to the most experienced stove operator. You can, if you have dump grates, shake all the ash and coal into the ash pit, then screen out the coal for reuse. Often it is better to remove the coal through the fire door without disturbing the ash layer. Leaving an ash layer will protect the grates and help support the new fire. Establish the fire again by following the procedure under Starting the Fire.

Burning Coke

Coke is similar to anthracite in its burning characteristics and most of the same procedures that apply to anthracite apply to coke. There are a few differences that need to be explained.

To make coke, coal—generally bituminous—is heated in an oven to remove the volatiles. As it is heated in the absence of air it swells, creating a porous structure. The by-products are recovered and used in many ways. What remains is mostly fixed carbon.

It is a relatively clean fuel. Occasionally it may be dusty if it becomes too dry. A light sprinkling with a hose before handling will reduce the dust.

Because the volatiles have been removed it is smokeless and doesn't create any soot accumulations. The chimney and stovepipe rarely have to be cleaned more than once a year.

Its porosity makes it a fuel that responds more quickly to changes in draft than anthracite, and it needs less attention between firings. The thermal efficiency of coke is equal to that of the hard coal. Because it weighs half as much as anthracite by volume, more frequent refuelings are necessary. You can think of the relationship of coke to hard coal as being similar to the relationship of softwood to hardwood.

Because it doesn't burn as long, coke is less desirable if you want the fire to burn all night or all day without attention. A larger firebox is needed. You can also extend the burn cycle some by reducing the drafts, using a finer grade, or tamping the top of the fuel bed slightly to reduce the air spaces.

Getting a coke fire started can be a little more difficult. With most of the volatiles already driven off, a higher temperature is needed to get the carbon to burn. This means that you need a good wood fire first. Once the coke is burning you can close the secondary air draft and control the fire with the primary air draft. Secondary air is not needed at this point. Adjustment of the dampers is more critical than with anthracite. With very few gases or flames visible it is harder to see when the damper is closed too much. A surface thermometer placed on the stove or stovepipe may aid you in getting the proper adjustment.

In recharging, even in mild weather when you don't need a lot of heat, it is best to fill the firebox. This keeps a more uniform fuel temperature. The amount of heat produced should be regulated with the dampers.

Burning Bituminous

Bituminous coal is mined in many sections of the United States. At least twenty-eight states have active mines. Its wide use in industry makes it more readily available in most areas. Although it is not as desirable a fuel as anthracite it is very reasonable in cost in some areas, especially where it is mined, and therefore is a good choice.

Because of its higher volatile content, bituminous is fired differently from anthracite. The *low-volatile* bituminous coals—those with a volatile content of less than 20 percent—are generally fired with the conical method.

The first fire can be made similar to the anthracite fire. Use paper, kindling, and wood to get a bed of coals established. Start adding the coal in layers, allowing each to ignite before adding more.

This coal burns differently from the anthracite. Because it has more volatiles, the first flames will be long, generally orange or yellow. There will probably be quite a bit of smoke, too. As the gases burn off, the flames become shorter and may change color slightly to correspond to the type of impurities present. The flame length also varies with the rate of burn, the longer flames indicating a hotter fire.

Form a Cone

Once the fire is established, add the coal to the center of the firebox, forming a cone. The larger pieces will roll down the pile to the outside, allowing more primary air to flow through, and creating a hot fire around the cone. This heat drives off the volatile gases, and the turbulence created increases the efficiency of the burn. After the volatiles are burned the coke formed will burn more slowly and you will get a long burn cycle.

Adjust the dampers about the same as for anthracite except allow more secondary air to enter and open the stack damper until the volatiles are burned. Before refiring, break up the cone a little with a poker, especially if it has caked over or formed a crust. Be careful not to mix the coal as this increases the chances of forming clinkers. When shaking the grates use short motions and stop when you see a glow in the ashes or the first red coals fall.

For overnight operation, shake the fire and add coal, forming the center cone. Allow the volatiles to burn off before closing the fire door and stack dampers. Close the ash pit damper to the desired heat level.

You will have more maintenance with bituminous coal. In handling,

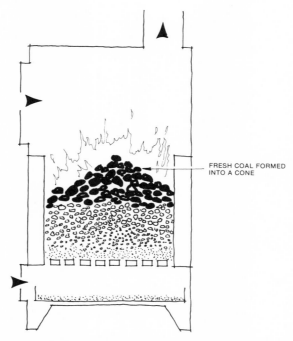

In firing low-volatile bituminous coal, a cone-shaped pile of coal is added.

there is more dust unless the coal was given a dustless treatment. Also, more soot will collect on heating surfaces and in pipes, requiring more frequent cleaning.

Burning High-Volatile Bituminous

High-volatile bituminous coal having more than 20 percent volatile content is easier to burn but gives off more smoke. It burns somewhat like wood in that it is easier to ignite and burns with longer, smoky flames.

To start a fire, shake the ashes and remove the clinkers to clean the openings in the grates. Open the dampers. Shovel two or three inches of fresh coal onto the top of the grate. Crumple up a few sheets of newspaper and place them on top of the coal. Add a few pieces of dry, finely split kindling and a few larger pieces. Light the paper and close the doors on the stove. When the coal is burning brightly start adding fresh coal in thin layers until the coal is up to the bottom of the fire door. Once the fire is well established, adjust the dampers, leaving the secondary air damper open a little so air can mix with the volatile gases.

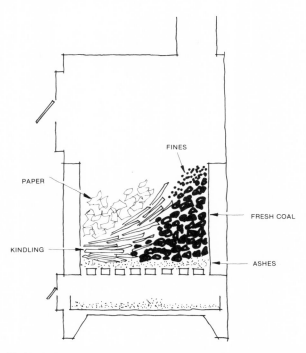

When kindling high-volatile bituminous coal, pile coal against one side of the firebox, and build a wood fire on the opposite side.

An alternate method for starting the fire is to heap up fresh coal against the back or one side of the firebox. Allow a little to cover the grate area. Place the newspaper and kindling against the sloping side of the coal and light the paper. This will ignite the pile from the outside and reduce the number of times that you have to add coal. When recharging, fill up the hole left where the kindling was, spreading the coal from side to side or front to back. Heat from the burning coal gradually penetrates the fresh coal, raises its temperature, and causes the gradual distillation of the volatiles. The hot coal causes a constant but slow burning of the combustibles and reduces the smoke to a minimum. It is important that any hot coals be shifted from the empty half before refueling, leaving just a layer of ashes. If not, you will have partial burning and a lot of smoke.

You can further reduce the smoke from this type of fire by covering the fresh charge with an inch or two of fine coal. Take this from the fines that accumulate at the bottom of the pile or purchase a few bags at the coal dealer. Adding the fines will keep the fire from spreading too rapidly and will force the gases into the hotter part of the fire. Before refiring, reduce

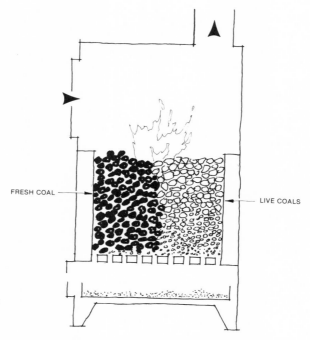

When recharging a high-volatile bituminous coal fire, place coal in the space left when the kindling burned.

the ash layer and break up the crust on the fuel side. The dampers are regulated the same as with other fuels except that more secondary air is allowed in to burn the volatiles coming off the top of the fire. More space should be left above the fuel when burning bituminous and other high-volatile coals to give a longer burning time before the gases reach the stovepipe. This will help to reduce the smoke.

Burning Subbituminous Coal and Lignite

These low-rank fuels are not commonly burned in domestic heaters except in areas near where they are mined. Subbituminous is found in Colorado, Wyoming, and Montana; lignite is found mainly in North Dakota and Texas. Their high moisture and ash content and low heat value make them less desirable than other types of coal. Overheating in storage can also be a problem. They should be stored in a tight bin with a cover.

Although standard stoves and heaters will burn these fuels without difficulty, several stoves have been designed specifically for them. Certain firing practices will help to increase their efficiency. Deep fuel beds are not necessary because the coal burns at a much lower temperature. These coals will burn well in converted wood stoves, those with a basket grate, and in fireplace grates because the fuel will remain kindled for many hours even while immersed in ash. Once started the fire does not die out as rapidly as it does with the harder coals.

More Ash, More Tending

With these fuels a greater amount of fine fly ash is generated and blown around in the firebox. Clean all flat surfaces in the stove regularly, otherwise heat loss may result. Also clean the stovepipe or the draft may be reduced.

A fire with subbituminous or lignite will have to be tended more often; otherwise the firing procedures are about the same as with the hard coals. More secondary air is needed to get the highest efficiency but very little smoke is generated if a good fire with a hot surface is maintained. If you have long, smoky, orange or yellow flames you are not getting fuel efficiency from the coal and you should try to operate a hotter fire.

Burning Coal in a Fireplace

Coal does not have the bright, cheerful blaze of wood but it does give a long, steady fire once established. To burn coal in the fireplace, get the proper grate, a *basket grate*. It should be rectangular with high side walls somewhat like a shoe box and have narrow spacing ($1/2$–$3/4$ inch) between the bars to keep the coal from falling out. Purchase the grate from a reputable dealer and get one of high quality made from good cast iron. Some grates imported from the Far East will not last through more than a few good coal fires. They will warp, sag, and burn out.

Remove any ashes from the fireplace and place the grate near the back. Twist a few sheets of newspaper and slide them under the grate. Lay a few pieces of dry finely split kindling on the grate and a few larger pieces above those. Allow air spaces between the pieces for the air and flames to go through.

Which Coal to Use

The type of coal to use will depend on what's available in your area and what your preference is. Try several types if possible. Cannel coal is one of the most popular coals for the fireplace, and most coal dealers carry it.

Caution: cannel coal should not be used in a closed stove. This is a bituminous-type coal with a high percentage of trapped oils and gases. It ignites quickly, puffs, sputters, and sparks quite a bit and burns hot. A good precaution is to close the fireplace screen when burning cannel coal.

Bituminous coal also works well in the fireplace. It ignites quickly, burns with a bright yellow or orange flame, and leaves a minimum of ash. Anthracite can be used but you will need a good bed of hot wood coals to get it started. Then add a few pieces of coal frequently to keep the fire up. Most of the combustion air should flow through the bed. This is difficult to achieve with a grate in the fireplace.

Open the fireplace damper before lighting the paper. In damp, foggy weather you may have to ignite a few pieces of paper on top of the grate or held up into the chimney to get a draft started up the chimney. Once the wood is burning brightly, place a few pieces of coal on top, being careful not to disturb the wood. Keep adding coal a little at a time, being sure that the bottom of the grate is covered. After the fire gets started reduce the damper opening to a point where smoke does not enter the room. Never close the damper when there is fire or live coals in the fireplace. If you do, poisonous smoke and carbon monoxide will be trapped in the house.

Avoid Losing Heat

Generally, a fireplace fire should be put out and the damper closed before you retire for the night. This way the chimney will not draw heat from your house all night. With the damper open the fireplace will remove the heat in a ten- by fifteen-foot room in about five minutes. That's why on freezing days you get very little warmth from a fireplace and on days when the temperature drops below 20° F. your furnace will run more than if the fireplace were closed. Adding glass doors to the fireplace will help some, but they also reduce the amount of heat that gets into the room. A better choice is to purchase a fireplace insert that has been designed to burn coal. These double- or triple-wall stoves that fit within or in front of the fireplace capture much of the heat that would have gone up the chimney and return it to the room.

Ash Disposal

A daily chore for coal stove owners is ash removal. After shaking the grate, let the dust settle for a few minutes before opening the ash door. Shovel the ashes into a metal pail with a cover and store it outside or in an area where there are no combustibles nearby.

Ashes often contain unburned coal and clinkers. You can recover some of the coal by screening. Make a square about 18 inches square using 1 × 3 furring strips. Attach a piece of ½ × ½ wire mesh to the bottom with staples. The mesh can be purchased at most hardware stores. Shake the ashes through this. It can be a dirty but worthwhile task.

The amount of ash you will get depends on the amount of impurities in the original coal. It can vary from 2 to 20 percent, with the highest percentages usually found in some forms of lignite. Coals used for domestic heating have from 8 to 12 percent and will give from seven to nine bushels per ton. Wood ash contains minerals that benefit the garden, but coal ash has some minerals that may be detrimental and so should not be used for that purpose. Usually the elements are part of some oxides or compounds. A list of some of those found is given in Table 10.

TABLE 10. ELEMENTS THAT CAN BE FOUND IN COAL ASH

Antimony	Chromium	Molybdenum
Arsenic	Copper	Nickel
Barium	Fluorine	Phosphorus
Bismuth	Germanium	Silver
Boron	Gold	Strontium
Bromine	Indium	Thorium
Cesium	Lead	Vanadium
Chlorine	Lithium	Zinc

Creosote

If you burn coal you don't have to worry about creosote and its inherent danger. Creosote is the tarry substance that forms inside the stovepipe and chimney flue from the condensation of the smoky moisture from burning wood.

During the early fall and late spring there are times when you need a little fire to take the "chill off the house." In the time that it takes to start a coal fire you could light a wood fire, warm up the house, and let the fire go out. Many homeowners have wood around for just this purpose. As we said before: you can burn wood in a coal stove but it's difficult and often unsafe to burn coal in a stove designed for wood. You may have to cut the wood a little shorter to fit some coal stoves and different settings on the dampers may be needed, but the wood will burn just as well.

If you burn more than a few wood fires each year you should be aware of how creosote is formed and its dangers. Green wood contains as much as 250 gallons of sap and water per standard cord of 128 cubic feet. If it's air dried for about a year and kept under cover it will still contain about 100 gallons. When the wood burns the water is turned to steam and carried out the chimney with the smoke. If the flue gases are cooled below 250° F. the moisture-smoke mixture condenses to a tarry liquid that sticks to the chimney and stovepipe. If at a later time you overheat the stove, the creosote can ignite, becoming a very hot and dangerous chimney fire. This fire burning at about 1500° F. can damage your chimney and ignite combustibles nearby. If you have a chimney fire, close the dampers on the stove, then call the fire department.

To reduce the amount of creosote that will form in your chimney use dry, well-seasoned wood. Also burn a hot fire rather than a smoldering one, allowing quite a bit of secondary air in through the fire door damper. This will burn the volatile gases in the smoke and give you more usable heat.

Stove Maintenance

The stove or heater should be kept in good repair at all times. This will result in better operation, higher efficiency, and a longer useful life. Some of the more important points to check are the following:

1. Remove the ashes daily. Too much ash in the ash pit will restrict the draft so the fire does not burn properly. It could also damage the grate by overheating it.

2. Clean soot and ash from heat-conducting surfaces when they start to build up. A 1/8-inch layer of soot can reduce heat transfer by as much as 10 percent. Also clean stovepipes and chimneys when the soot gets thicker than ¼ inch. A wire brush works well.

3. All doors and dampers should be examined occasionally to see that they fit tightly and are operating properly.

4. Any cracks between sections of castings, stovepipe connections, or other locations should be sealed with furnace cement. Draft leaks will impair the operation of the stove and can be found by slowly moving a lighted candle near the joints. The flame will be drawn toward any opening. It's best to do this when you have a fire going.

Your Stove During the Summer

At the end of the heating season your stove should be cleaned, inspected, and repaired so that it will be ready for next fall. Ashes should be removed so that they don't corrode the metal parts. If your stove is not in the way, you can leave it in place, but you may want to move it to a dry storage location in the garage or basement. The following are a few of the maintenance chores that you should do at this time.

1. Disconnect the stovepipe and clean it. Clean the chimney, if necessary, then remove the soot and ashes from the clean-out at the bottom. If you don't have the equipment or don't want to climb on top of the roof, hire a chimney sweep. He has the equipment and know-how to do a good job.

2. Remove the ashes, soot, and clinkers from the fire chamber and ash pit. A wire brush and old vacuum cleaner are good tools for this job. Don't use your best vacuum cleaner, since this task is wearing on the parts.

3. Coat the metal surfaces inside with a light oil or silicone spray for rust protection.

4. Use stove polish or high-temperature, touch-up paint on external surfaces to keep the stove from rusting.

5. Look for broken or warped grates, doors, or dampers and order replacement parts.

6. Check for cracks between castings where furnace cement or asbestos caulking has been loosened. Replace as necessary.

7. Place an open coffee can of silica gel inside the firebox to absorb moisture. This gel can be purchased at any drug store.

8. If your chimney doesn't have a cap on it, fit a piece of sheet metal so that it covers the top to keep the rain out. A stone or piece of steel suspended from the center of the cover inside the chimney will hold it in place.

9. If you remove the stove and the stovepipe, close the thimble with a thimble cover or piece of sheet metal.

Problem Stoves

You can read this book and understand all about how the stove operates, but you really don't learn how to fire the stove until you do it. Each stove installation is a little different, and only through experience do you become familiar with the proper technique for your own stove. There are times when problems arise with no obvious answers. Following are a few of the more common ones and some suggestions on what might be the cause and remedy.

Can't Get the Fire Started

1. Kindling will not burn

- Use finely split pieces of dry wood. Softwoods (pine, spruce, hemlock, cedar) work best because of the resins they contain. The wood should be dried at least a year and be stored under cover.

2. Chimney not drawing

- Are the chimney and stove dampers open?
- Is the chimney blocked?
- Heat chimney and get the draft started by lighting some newspaper on top of the kindling.
- Are the ashes in the ash pit blocking the grate?
- If restarting a fire, no more than one inch of ashes should be left on top of the grate.

3. Wrong coal

- Use recommended type for stove. Anthracite is harder to start than other types and requires a good bed of wood coals. Sometimes the use of a small amount of coal briquettes, usually a soft coal, helps to get the fire started more quickly.
- The coal is too large. It takes more heat to get larger pieces started.
- The wrong firing method is being used. Refer to the section on Firing for the correct method for your coal.
- Too much coal at the start blocks the draft. Try adding a few pieces at a time.

Low Heat Output

1. Insufficient draft

- Soot build-up in stovepipe or chimney.
- Air leaks around stovepipe, thimble, or clean-out door. Use furnace cement to seal.
- Chimney not tall enough. Minimum height should be fifteen feet. Twenty feet is better.
- Chimney too small. 8 × 8 flue is minimum size for most stoves; 8 × 12 inch is better.
- Chimney too large. Use a cone on top or a separate stainless steel pipe within the chimney.

2. Stove not adjusted properly

- Too much secondary air. A small amount of cool air can curb a fire.
- Stove damper closed too much.
- Clinkers blocking grate.
- Ash build-up in corners of grate.
- Too much ash in pit.
- Air leaks in stove around joints, and these should be sealed with furnace cement.

3. Poor heat transfer

- Soot on heat transfer surfaces inside stove. Clean with a brush.
- Poor heat movement within the room or house. Use small fan, floor registers, or duct system.

4. Fuel

- Not enough depth of fuel in firebox.

Smoke Smell in the House

1. Chimney related

- Stovepipe damper closed. Open slightly.
- Chimney or stovepipe partially blocked with soot. Set up regular maintenance schedule to check this.
- Down drafts in chimney. Can be caused by nearby trees, a taller section of the house, or other building. Use chimney cap or extend the height of chimney one section.

2. *Stove related*

- Smoke entering house when fire door is open to refuel.
- Leaks in stove.
- Puff backs when a mixture of combustibles, gases, and air are ignited from flame traveling through fuel bed.

3. *House related*

- House too tight. Bring in make-up air by opening basement door or installing separate pipe to outside.

Coal in Ashes

1. *Operator related*

- Too vigorous shaking. Shake only until first red coals drop into ash pit.
- Grate not positioned properly. Be sure there are no large cracks between grate sections.
- Coal too small. Use proper size for the stove.

2. *Stove related*

- Broken grate. Repair parts may be available through stove shop or manufacturer.

Burns Too Much Coal

1. *Fuel related*

- Poor coal. Purchase coal with a low ash content from a reputable dealer.
- Wrong size. Use size recommended for the stove.

2. *Operator related*

- Too much secondary air. Close the damper.
- Too much draft. Close stovepipe damper.

CHAPTER 5

CENTRAL HEAT

When should you consider installing a central heating system rather than a stove or heater? This question is often asked if you are building a new home or considering replacing an old furnace. There are many factors to consider before making this decision. Before discussing some of these, let's define what we are talking about.

Equipment Definitions

A *furnace* differs from a stove or heater in that a sheet-metal jacket is used to enclose the firebox to capture the heated air and direct it through ducts to the rooms in the house. It may have a blower to distribute the heated air or it may depend on gravity flow. It is generally large enough to heat the whole house. In most furnaces the firing rate is controlled automatically by a thermostatically controlled blower or damper. Other safety devices may be installed to prevent overheating.

A *boiler* is basically the same as a furnace except that the fire heats water which is distributed through a pipe system. Control is different in a boiler in that the temperature of the water in the jacket or pipes surrounding the firebox is maintained at a pre-set temperature, generally 180° F., by the firing rate.

The temperature in the house as sensed by the thermostat is maintained by circulating the heated water through the radiators. Because of the large amount of water in radiator-piping systems, a boiler having a large water capacity will maintain a more uniform house temperature.

The *add-on furnace* or *boiler* is a solid fuel-burning device that is added or attached to an existing central heating system to supplement that system. Many coal furnaces and boilers can be installed as add-on units.

The advantage to this type of unit is that the cost is greatly reduced because the existing distribution system is used. Some additional insulation or protection may have to be provided around the ducts or pipes to handle the potentially higher temperatures.

A *multifuel furnace* or *boiler* will burn coal and oil, gas, or wood. Some units have separate fireboxes for burning the different fuels; others use only one firebox. Combination units with fireboxes designed for each type of fuel are usually more efficient than units with a single firebox. For example, a firebox for burning coal would be too large for the proper operation of an oil-fired burner. Another factor to consider is the potential fouling of the gas or oil burner by the soot, even in the dual firebox design. Except for the design of the firebox these units are similar to other solid fuel-burning units in operation and control.

How to Decide

If you wish to heat the whole house a furnace or boiler makes sense. Moving heat through ducts or pipes is more efficient than using floor registers or doorways. This will also produce a more uniform temperature throughout the house.

A stove in a small, well-insulated house can do a good job of providing heat but if your house is large and of an older vintage a central system is a better choice.

A central heating system will cost two to four times as much as a stove. This is a major investment, often in the price range of a new car. But the government may help you out on the cost; find out what tax incentives are available from the state or federal government. For more information call your state energy office, listed in the Appendix.

The central heating system is usually in the basement. This keeps the ducts and pipes away from the living area. It also keeps more of the dust and dirt associated with burning coal in the basement.

There is generally less work operating a furnace or boiler than a stove, particularly if you have a stoker that will feed the coal automatically. Large ash pits have to be cleaned less frequently and the larger firebox holds a larger charge.

More on Market

You may have difficulty finding a furnace or boiler that will burn coal, although the number of them on the market is increasing rapidly, as new manufacturers join several old-line companies. Some manufacturers of wood stoves and furnaces are developing coal-burning units. If you have an idea which model you like, check with the owner of one to see how it works. It usually takes several years to work the bugs out of a new design, so it's best not to do the testing for the manufacturer. Another source of proven designs is Europe. Several companies import European models and distribute them through a network of dealers, generally the stove shops. See the listings at the end of this book.

Safety is important. An approved furnace or boiler properly installed is safer than a stove or heater. Most states require that the unit be installed by a licensed technician just as it would for an oil or gas unit. Installation of the proper safety devices can make the unit as safe as any gas or oil unit.

In sizing a solid fuel unit for a new home the builder or heating contractor should calculate the size of a conventional furnace needed, then subtract 10 to 15 percent. Coal furnaces and boilers can be overfired to handle the few very cold nights during the winter but it's difficult to damp down the fire to provide the small amount of heat needed in the late spring and early fall if it is too large. Some manufacturers recommend using an alternate heat source (small stove, oil or gas on a combination unit) when the temperature outside is above 40° F.

Furnace Design

The furnace differs from a stove in that the heat is moved through ducts to the different rooms. Additional controls and safety devices are used to make operation easier.

The illustration on page 126 shows the basic parts of a furnace. It is usually made from cast iron or steel and contains a firebox, heat exchanger, and metal enclosure. Generally the materials that are used in stoves are used in the construction of the furnace. The firebox is lined with either cast iron or firebrick to protect the metal from high temperatures. In a furnace these may get up to 3000° F. As with stoves, a full shaker grate system is the best for easy operation. A large pit or pan will reduce the frequency of cleanings to about once a week.

Draft System

The draft system will vary with the type of furnace. Older-style furnaces which you may still see in homes of the 1930s era or which can still be purchased from a few manufacturers usually have manual controls. In some systems the damper on the ash pit door and the check damper in the stovepipe are operated by moving a lever in the room above the furnace. The lever and the dampers are connected by a chain. To get more heat

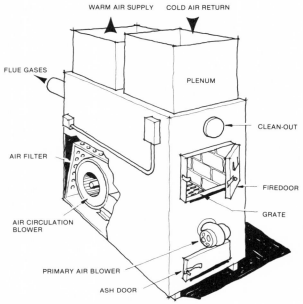

The basic parts of a furnace

you adjust the lever so that the bottom damper opens more and the check damper closes. A more modern version of this is to use a thermostatically controlled damper motor to make the adjustment automatically. The thermostat is in the living area. A more positive draft system includes a small blower attached to the ash pit door. When the thermostat calls for heat and the temperature of the air in the furnace is below a set point, usually 100°–120° F., the blower activates and forces air through the fire, increasing its intensity. This type draft system can also be operated manually during a power failure by using the draft control on the ash pit door.

Secondary Air

In most of these automated systems the secondary air above the fire is controlled manually. Because coal does not have as many volatiles as wood, adjusting the secondary draft so that it allows a small amount of air continuously will be all that is necessary.

Most furnace installations require a separate class A chimney flue. This can be either masonry or an approved all-fuel factory-built metal chimney. The installation should meet the manufacturer's installation specifications and accepted practices for conventional oil- or gas-fired units.

The better furnaces use a separate heat exchanger to capture more of the heat from the gases before they enter the chimney. This steel box, sometimes finned and sometimes baffled inside, transfers the heat from the gases to the air that is circulated throughout the house. The heat exchanger should have a clean-out door to remove any fly ash deposited.

The furnace has connections for ducts which carry the heat to the various rooms in the house. Gravity furnaces will move the air without blowers. As the air is heated it becomes lighter and rises through the ducts and enters the room through a register. The cooler air returns to the furnace through a return-air duct. The furnace must be located below the lowest floor to be heated.

The registers are placed near the floor, usually in the baseboard. Return-air grills are located in the floor. Ducts should be insulated at least in areas that you don't want heated such as a crawl space or garage.

Best Location in Center

The ideal location for the furnace is under the center of the house. This allows all ducts to be about the same length and makes for more uniform temperature throughout the house. A gravity furnace is economical to install, and, because it doesn't have any blower, economical to operate. It responds rapidly to heating needs even in cold weather.

A forced-air system uses a squirrel cage blower to move the air through the ducts. With this system you are not limited to a central location, as each register can be adjusted to give a balanced flow of air. A filter can be used to remove dust from the air. Registers can be placed in the baseboard or near the ceiling, giving more freedom in the placement of furniture. Grills for return air can be located in the floor or baseboard, generally below windows or other cold spots.

Humidifier

A humidifier can be incorporated into most furnaces to add moisture to the air. Increased humidity is desirable during the winter to increase the comfort level at a lower temperature, reduce respiratory problems, and keep furniture from drying out. A humidity indicator should be placed in the living area.

On some furnace units the blower can be switched on manually for summer operation. Although it will not cool the house a great deal it will provide air movement which speeds up evaporation of body moisture and makes you feel more comfortable.

Most furnaces can be used either as a separate heating unit or as an add-on to an existing heating system. If you are installing a separate unit you will have to add the cost of the ducting in your estimate. This can be as much as half the total cost of the installation. An add-on unit uses the existing furnace ducting and the only additional expense is the connections between the two units.

Furnace Installation

Many of the set-up procedures and safety considerations that apply to coal stoves also apply to coal furnaces. A review of the section on installation should be made before you begin. Just as important is to study the manufacturer's instruction manual. Hooking up a coal furnace requires a good understanding of heating systems and electricity. Because most homeowners do not have this specialized training it's best to have a licensed heating contractor or technician do the installation. In some states this is mandatory. Check with your local building official or fire marshal before you start.

Special attention should be given to several areas of the furnace installation. Let's look at these in a little more detail.

1. The temperatures developed in the chimney from a coal furnace can run several hundred degrees higher than with a stove because of the greater amount of fuel being burned. Use only a tile-lined chimney that's in good repair or an all-fuel factory-built metal chimney. A separate Class A flue is required for most separate or add-on units.

2. Follow the manufacturer's recommendations for clearances from combustible material. You will find that these clearances are generally less than for a stove installation because the heat exchanger, enclosure, and insulation allow less heat to escape through the side wall and more to be carried upstairs. If you are not setting the furnace on a concrete floor, a good floor protection is to use four-inch tile or concrete blocks laid so that air can circulate through them.

3. Increased distances from floor joists, flooring, and other combustibles are required for the ducting as compared to an oil- or gas-fired unit. The possible higher air temperature and less positive fire control make this necessary. NFPA minimum distances for coal or wood furnaces, add-on units, and dual fire units are shown in the illustration. If you are using an existing duct system you will probably have to increase the

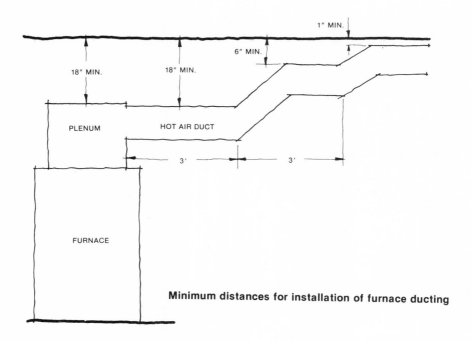

Minimum distances for installation of furnace ducting

clearance, especially where the duct goes through the floor or ceiling. If you can, cut out the extra wood. If not, insulation may be necessary. The local building official can advise you on these special situations.

4. Add-on units must be hooked up properly to be safe. Although there are usually several ways that the connection can be made between the two units, the preferred one is with the coal furnace being placed in parallel with the oil or gas unit. Series connections with the coal furnace being placed after the conventional furnace will work but only if the coal furnace is large and has adequate-size ducts, otherwise back pressure will develop, causing overheating and possible blower motor burnout.

5. Controls are needed to prevent overheating. This can be caused by such things as a damper being inadvertently left open, an exceptionally strong draw on the chimney, or a stuck damper control. When a stove overheats it just radiates more heat into the room. With a furnace having insulated walls, the heat is transmitted through the ducts. This could cause overheating of nearby combustibles. Protection devices include a combustion air control that closes the primary draft, and a plenum temperature limit control that operates the blower continuously when the temperature exceeds 200° F.

130 HEATING WITH COAL

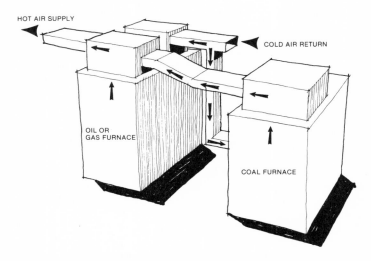

Installation of an add-on unit, with the coal furnace placed in parallel with the oil or gas unit

6. In the event of power failure, the blower and the controls become inoperative, and temperatures can build rapidly, causing possible damage to the furnace. An electrically operated primary air control that fails safe is better than a mechanical thermostat.

Boiler Design

If you want steady, even heat in your home, even on the third floor, a hot water boiler with zone circulation may be the best choice. Although more houses are heated with furnaces in the United States a boiler system has several advantages.

1. Water is a better heat storage and transfer medium. It holds the heat longer than air, giving an extended heating period even after the circulating pump shuts off.

2. The piping takes up less space than a duct system. This is a particular advantage in a basement with a low ceiling.

3. A tankless coil is available with some boilers that will heat domestic hot water, eliminating installation of a separate heater.

A boiler differs from a furnace in that it contains a water jacket and pump instead of heat exchanger and blower. The water jacket surrounds the firebox. The gases from the fire are exhausted through horizontal or vertical tubes that pass through the jacket. In a well-designed unit of 100,000 Btu per hour output, about the size used in most homes, the jacket will hold more than fifty gallons of water. This capacity is needed to fill the radiators with hot water when the thermostat calls for heat and the circulating pump starts.

Most boiler manufacturers use steel plate construction rather than cast iron. The firebox may be lined with firebrick and the doors and grate made from cast iron. ASME certification, which is available with some units, means that the material quality, welding, inspection, and pressure testing have met a set of standards acceptable to building officials throughout the country.

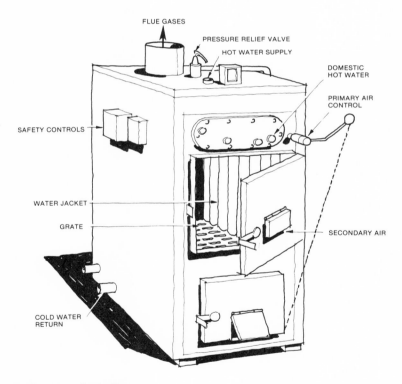

The basic parts of a boiler

Different Controls

Control of a boiler is different than a furnace. A heat-sensing aquastat is often used to adjust the damper so that water temperature in the water jacket is maintained at a preset level, usually 165°–180° F. A second heat-sensing thermostat is located in the living area. This monitors the room temperature and turns on the circulating pump. The pump takes the hot water from the jacket and moves it through the radiators where the heat is used to warm the room to the temperature set on the thermostat. This system is similar to a conventional oil- or gas-fired boiler installation.

In older houses you may still find a gravity hot water system. Water heated in the boiler rises in the supply pipes and the cool water from the radiators returns to the bottom of the boiler. No pumps are used. It is very difficult to adapt this system to a coal-fired unit and get good temperature control. It would be better to have it converted to a forced system with a pump.

If you still have a steam heating system and are interested in evaluating a conversion to coal you are in luck. At least one company (Hardin Manufacturing Co.; see catalog section) makes a boiler that can be used with a steam system. Steam is not used in new homes today because it is not as responsive to heat demands as hot water. It is also difficult to find a heating contractor who knows how to install a new steam system.

Just as with furnaces, most boilers can be installed as an add-on unit to an existing oil or gas boiler. Use of the existing pump, radiators, and piping system can reduce the cost of the installation considerably.

Boiler Installation

Although coal-fired boilers are made to fit through a standard door, you will need plenty of help to get one into your home, especially if it has to go through a bulkhead. Most units weigh 600 to 800 pounds and several weigh over 1,000 pounds.

A boiler installation is potentially more dangerous than a furnace because as the water is heated it expands slightly, and even more if it boils and turns to steam. If provisions have not been made to keep the water temperature below boiling and to absorb the expansion of the water a dangerous explosion can result. For this reason most manufacturers and building codes require that a boiler be installed by a licensed technician.

This makes good sense. Why subject your family and yourself to this potential danger?

Installation of a straight coal boiler system is similar to a conventional oil or gas unit except for a few components. Several overheat prevention devices are used.

1. The *aquastat* that senses the temperature of the water in the jacket adjusts the primary draft control and closes it if the water exceeds a preset point, usually 180° F.

2. A *heat sensor* in the jacket wired in parallel with the circulating pump activates the pump to operate continuously if the water temperature exceeds 200° F. This moves the hot water through the radiators where it is cooled.

3. A conventional *pressure-temperature relief valve* vents water or steam out of the system if either exceeds the set point. For boilers the rating should be thirty pounds per square inch (psi) and 200° F. Don't use the 75 or 150 psi rated relief valves designed for domestic hot water heaters.

4. A few manufacturers use a *fusible plug* or *temperature-operated nozzle* that activates to douse the fire. Overheating should be avoided as this can seriously damage the unit.

The boiler should be of airtight construction so the fire can be kept under control even on very windy days. Tight construction is also needed for high efficiency. Air leaks around the doors and at the joints allow excess air into the firebox. This cools the fire and carries heat away from the water jacket area. Check your boiler several times a year at night and when the fire is out by placing a lighted flashlight inside the firebox. Look for light leaks around doors and closed dampers.

Add-On Boilers

Add-on boilers can be installed either in series or in parallel with the conventional boiler. The series installation as illustrated generally costs less because the existing circulating pump is used to move the water through both units. A disadvantage to this system is that the water cools as it moves through the boiler that is not operating. For example, if the coal boiler were being fired the water would still have to travel through the oil unit and be cooled before it reached the radiators.

In a parallel installation two circulating pumps are needed. Water to be heated can then be circulated through either boiler independent of the

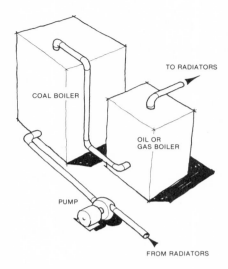

Add-on boiler installed in series with a conventional boiler

other. This type of installation requires additional valves and proper thermostat connections and should be done only by a competent heating contractor familiar with add-on units.

Multifuel Furnaces and Boilers

Several manufacturers make units that will burn a solid fuel such as coal, and gas or oil. Both single and separate firebox designs are available. In single firebox construction the oil or gas is often used to ignite the wood or coal. Its disadvantage is that the gas or oil unit is subject to fouling from soot and creosote from the solid fuel. In the separate firebox design the chambers are sized for the fuel used. The gas or oil unit is more efficiently burned in a smaller firebox. A larger firebox is needed for the coal. Generally the two chambers are connected by a passageway and a single chimney connector is used. In both types of units there is the possibility of small explosions should the oil or gas valves malfunction or leak. The vapors could be ignited by heat or live coals in the solid fuel section.

At least one manufacturer has developed a boiler with two independent combustion systems. Both are surrounded by water and enclosed within the water jacket. In all units the control will automatically switch to gas or oil should the coal fire burn out.

Multifuel units fit in nicely for new house construction or where a furnace or boiler is being replaced. Although the initial investment is greater than for a single-fuel unit, the advantage of having the back-up for weekends when you are away often pays. It also works well on the gas or oil in the spring and fall when you don't want to start a coal fire to take the chill off the house. The gas or oil unit in a boiler can also be used to heat water during the summer.

Automatic Coal Burners

One step above a manual-feed boiler is the magazine-type unit. A large container is built into the upper section above the grate. This holds one or more day's supply of coal. Coal is fed into the firebox as needed. (See catalog listing in the back of the book.)

Almost complete automation is available with a *stoker*. This feeder-burner unit can be attached to some makes of both furnaces and boilers. The major components of the most common stoker are the hopper, blower, coal screws, and retort.

The *hopper* is a large bin or box that will hold 300–400 pounds of coal and needs to be filled about once a week. A second type unit will feed directly from a bin, eliminating one handling. Stoker units are available that will burn most types of coal. You will have to purchase a much

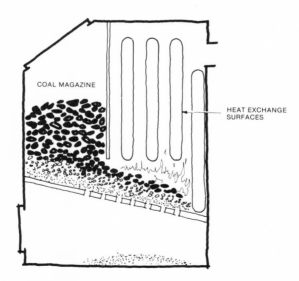

Magazine unit, with coal supply held at left

136 HEATING WITH COAL

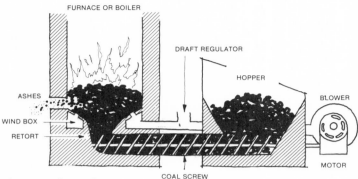

The basic parts of a stoker

smaller size coal so that it can be handled by the auger. Usually this is buckwheat or rice-size anthracite and stoker-size bituminous. Before purchasing a stoker find out what size coal is required and if it is available from your local dealer.

Auger Moves Coal

Located in the bottom of the hopper or bin is a *steel auger*, usually three or four inches in diameter. It is powered by a small gear motor and operates within a tube to carry the coal to the retort. The speed and therefore the rate of feed is adjustable and controlled by the amount of heat needed in the home as sensed by the thermostat upstairs. Surrounding the feed tube is a second tube that carries air from a blower to the *retort*. This is the primary air for combustion. The amount of air is also regulated to coincide with the feed rate.

Converting to Automatic

To convert a furnace or boiler, the firebox must be modified by removing the grate and ashpan and inserting the retort. The retort holds the coal while it is being burned and must be designed differently for each size and type of coal. The retort is shaped like the horn on an old Victrola phonograph. The coal being forced in by the coal screw burns on the top and the ashes fall off into an ash box. A few stoker units are so designed that the ashes are removed mechanically and placed in a covered metal container.

Correct installation is necessary to get proper operation. The unit should be installed according to the manufacturer's recommendations by a knowledgeable technician. Most units can be placed into the furnace or boiler through the side or rear but room should be left to remove it for servicing or cleaning.

The control system for a coal stoker is similar to that for an oil burner except that a timer operates the stoker for a few minutes every half hour or hour to keep the fire burning. A heat-sensing stack thermostat may also be used to keep the screw from filling the firebox with coal should the fire go out.

Operation

To start a fire, operate the screw until coal nearly fills the retort. Cover the coal with sawdust or wood chips and ignite with a newspaper or torch. Adjust the air flow and when the coal is burning across the top start the screw motor on the slowest speed. Adjust as the fire picks up.

Stokers require a minimum of care. Those used in home heating systems are smaller versions of industrial units and tend to be overdesigned. The following are a few items that will need attention.

- Occasionally oiling the bearings and gear motor.
- Cleaning the blower or fan unit.
- Removing clinkers.
- Using the recommended size and type of coal.
- At the end of the heating season thoroughly cleaning the stoker and coating the surfaces with oil.

Should You Buy a Stoker?

Now that you know what a stoker is and how it works should you invest in one? Here are some factors that will help you make that decision.

1. Do you have the room for a stoker? It usually takes an extra three to four feet to one side or in back of the furnace or boiler.

2. Is a competent installer available? Stoker installations have not been very common in most sections of the country during the past few years.

3. Are service and parts available? Although stokers are generally trouble-free, occasionally they will need a replacement part and they require an annual cleaning and adjustment.

4. A stoker reduces the amount of dust and ash that gets into the living area. Also less attention is needed for firing.

5. Most stoker units will provide more uniform heat than a manual furnace or boiler and may even increase the efficiency of the heating unit.

6. Buying and installing a stoker unit will increase the cost of a furnace or boiler about 50 percent.

Draft Requirements

As with a coal stove, the available draft and its control are important in the operation of the furnace or boiler. Because you are burning more fuel per hour and have a larger grate area, more draw on the chimney is needed. Some manufacturers list the minimum draft requirements for their units, others do not. Most require at least 0.05 inches water gauge to operate properly. Chimney height, cross-sectional area, and the temperature of the flue gases affect this.

Two other factors that influence the draft are the *size* of coal being used and the *rate* at which the coal is being burned. Increasing the coal one size will often allow you to increase the burn rate with the same amount of draft. This occurs because there is more space between the pieces and therefore less resistance. The stove dealer or furnace installer usually has a draft gauge to make this measurement.

If you don't have adequate draft, purchasing a unit with a forced draft, usually a smaller blower, can solve the problem. If you have too much, and this is common in multistory houses with a large chimney, you

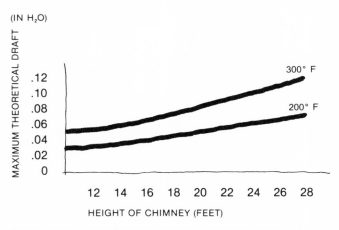

Temperature and height of chimney affect the draft.

will need an automatic barometric damper in the stovepipe. This allows the chimney to draw some of its air from the basement rather than through the furnace draft system.

Hand Firing

The same methods are used to fire stoves, furnaces, and boilers. Review Chapter 4 for the methods to use with the type of coal being fired. There are several differences.

1. The depth of the fuel bed is important as it regulates the flow of air through the fire. With a stove the location of the door above the grate limits the depth of the bed you establish. With the larger central units there is generally more latitude. This can work to your advantage if you lack an adequate draft. A deep bed can be maintained on windy, cold days when the draft is good and a shallower bed used on cloudy, damp days when the draft is poor.

2. A larger size of coal is used with the larger units. Follow the manufacturer's recommendations. Usually nut or stove anthracite and nut or egg bituminous are used.

3. With the larger ash pit less frequent cleaning is necessary. Don't get complacent, though, and allow the ash to build up too high. Too much ash could damage the grates.

4. In mild weather you may have to keep a deeper ash layer (two to three inches) or even place a thin layer of ash on top of a new charge of coal.

5. Open the dampers the minimum distance necessary to obtain the desired heat and maximum efficiency.

6. If you are burning bituminous coal, purchase the noncaking type. Caking will cause the coal in the fuel bed to become fused into a solid mass which must be broken up about one hour after firing. Care should be taken so that the ash is not mixed with the fire, or clinkers will form.

Maintenance

Although central heating systems require very little attention they usually get less because they will continue to function even under adverse conditions. An annual cleaning and check-up by a qualified serviceman is a must. Just as with an oil or gas unit the inspection will pay for itself in reduced fuel usage. It will include removal of the soot from the heating surfaces, lubrication of moving parts, adjustment of the controls, and checking the stoker if your furnace is so equipped.

During the heating season you can maintain the system. Accumulations of soot and scale will cause a portion of the heat produced in the firebox to pass through the furnace instead of being absorbed by the air or water. Clean-out doors are provided in strategic places so that you can insert a wire brush or vacuum nozzle. A special brush may be needed for a unit having an odd-shaped heat exchanger or fire tube water jacket. These usually come with the unit or are available from the dealer or manufacturer.

You should draw off a little water from boilers to remove any dirt or sludge that may have accumulated. Also see that all pipe fittings are tight.

Clean the heat transfer surfaces upstairs. Dust and dirt accumulated on radiators and floor registers can block heat movement and cause dirt streaks on the walls.

At the end of the heating season clean the furnace thoroughly, then spray the surfaces with a light lubricating oil. The moisture in the basement combined with the deposits in the furnace create an acid which corrodes metal surfaces. If left unattended, this can cause leaks in pipes and heat transfer surfaces.

Converting a Converted Furnace Back to Coal

In the 1930s and 40s when oil and gas were gaining in popularity many coal furnaces were converted to the modern new fuels. For oil this conversion consisted of:

1. Removing the grate and the ash pit door.

2. Reducing the size of the combustion chamber by lining it with refractory firebrick.

3. Inserting the oil burner and then sealing with firebrick the front of the ash pit door through which the burner entered.

4. Sometimes adding this firebrick in the area of the heat exchanger to increase turbulence and therefore heat transfer.

5. Changing the control system.

6. Adding a barometric damper to reduce the draft.

Oil-fired boilers were converted with essentially the same procedure. The gas-fired conversion burners usually came with their own refractory chamber so the firebox was not modified. These conversions often did not burn the fuel very efficiently. I have measured some that were as low as 55 percent efficient.

There has been interest by some homeowners in converting back to coal or wood. What is involved?

First, see if you can find the grates, ash pit door, and controls. If you can you're in luck. If not, determine who manufactured the furnace or boiler, not the burner. Look for a nameplate or a label on one of the castings. If you can locate the name of the manufacturer, you can find out if they are still in business. You might try checking with local heating contractors or the parts companies listed in Chapter 2.

Second, if you have the parts or can locate them, see if you can find a heating contractor who will do the conversion. Someone who made the original conversions may be best. The burner will have to be removed, firebrick chipped out, and the unit inspected for damage. You may also want to have the furnace or boiler modernized to include a new control system and safety devices.

Third, if the manufacturer of the original unit can't be located you may be able to adapt another grate system. You should expect to spend $200 minimum for a conversion and maybe much more. If on inspection you find that your unit is deteriorated and nearly burned out it may be better to put the money toward a new system.

CONCLUDING THOUGHTS

If after reading this book you have concluded that burning coal is a lot of work, you are right. It is. It's much easier to adjust the thermostat of the oil or gas furnace.

If you now feel that you can master the working of a coal stove, you are right. You can. The operation of a stove requires special skills that develop with experience just like driving a car through a congested city or cooking a holiday meal.

If you have perceived that the use of coal in your home will help the energy crisis and balance of payments, you are right. It will. For every ton burned 150 fewer gallons of oil have to be imported.

If you now think that by burning coal you will become rich, forget it. You won't. The price of coal will rise just like oil or gas, only maybe not quite so fast.

If you sensed that using more coal will increase the amount of pollution, you may be right. Only time will tell. It's true that all fuels cause pollution. Just how much and what kind varies widely.

If you think that a coal stove will try your patience, you are right again. It will, especially if your fire goes out on a cold night and you have to shake it down and start all over.

If you envision the supply of coal will last many years, you're right. Adequate supplies will still be around many years after you have passed on to greener pastures.

If you have concluded that by putting a coal stove in your home you can still have the comfort and security that you now enjoy, you are right. A well-planned stove set-up, safely installed and operated, can achieve all this.

I hope that by reading this book you have learned that installing a stove is not the right thing for everyone to do. It is one of several alternative heating systems that we have at this time. You should evaluate all of these carefully before you make a decision.

REFERENCES

Baker, L.D. et al. *Burning Wood*. Ithaca, New York: Northeast Regional Agricultural Engineering Service, Cornell University, 1977.

Brame, J.S.S. & King, J.G. *Fuel: Solid, Liquid and Gaseous*. London: Edward Arnold & Co., 1935.

Fellows, J.R. *Fuels and Burners*. Urbana, Illinois: Small Homes Council, University of Illinois, Circular G3.5, 1953.

Garver, H.L. *Your Farmhouse: Heating*. Washington, D.C.: USDA, Miscellaneous Publication No. 689, 1950.

Graham, F.C. *Audel's Engineers and Mechanics Guide* 6. New York: Theo. Audel & Co., 1937.

Haney, J.W. *Comfort and Economy in Heating the Home Using Coal or Coke as Fuel*. Lincoln, Nebraska: Engineering Experiment Station, University of Nebraska, 1932.

Israel, G.H. *Sales and Service Manual: Old Company's Lehigh*. Lansford, Pennsylvania: Lehigh Navigation Coal Co. Inc., 1949.

Johnson, A.J. and Auth, G.H. *Fuels and Combustion Handbook*. New York: McGraw-Hill Book Co., 1951.

Kristia Associates. *A Resource Book on the Art of Heating with Wood*. Portland, Maine: 1976.

Landers, W.S. & Parry, V.F. *Domestic Storage of Bituminous Lump Coal and Its Performance in a Hand-Fired Furnace*. U.S. Bureau of Mines, Report of Investigation, No. 3759, 1944.

Lytle, R.J. & Lytle, M.J. *Book of Successful Fireplaces.* Farmington, Michigan: Structures Publishing Co., 1977.

Nicholls, P. et al. *Five Hundred Tests of Various Coals in House: Heating Boilers.* U.S. Bureau of Mines, Bulletin 276, 1928.

Nicholls, P. & Landry, B.A. *Coke as a Domestic Heating Fuel.* U.S. Bureau of Mines, Report of Investigation No. 2980.

Oppelt, W.H. et al. *Combustion of North Dakota Lignite in Domestic Heaters.* U.S. Bureau of Mines, Report of Investigation No. 6581, 1960.

Palmer, E.L. & Bartok, J.W. *Burning Coal.* Ithaca, New York: Regional Agricultural Engineering Service, Cornell University, Bulletin FS-11, 1978.

Schmidt, R.A. *Coal in America.* New York: McGraw-Hill Book Co., 1979.

Shelton, J.W. *Wood Heat Safety.* Charlotte, Vermont: Garden Way Publishing, 1979.

Steiner, K. *Fuels and Fuel Burners.* New York: McGraw-Hill Book Co., 1946.

Twitchell, M. *Wood Energy: A Practical Guide to Heating with Wood.* Charlotte, Vermont: Garden Way Publishing, 1978.

GLOSSARY

Add-on furnace or boiler. A solid fuel–burning heating unit that is connected to an existing central heating system. The distribution of heat is through the existing hot air ducts or pipe system.

Anthracite. A "hard" coal having very little dust, smoke, or volatiles. It contains more than 90 percent carbon and is more difficult to ignite than other types of coal.

Approved. A stove that has been accepted for installation by the state or local building official or fire marshal.

Aquastat. A heat-sensing device placed in the water jacket of a boiler. It is used to control the rate of fire and thereby the temperature of the water.

Asbestos millboard. Soft asbestos sheet material used as a noncombustible material. Because of potential health damage from breathing the asbestos fibers, other acceptable materials such as sheet metal should be used if possible.

Ash. The unburnable mineral part of coal or wood. The amounts in different types of coal or wood vary widely.

Back puffing. The emission of smoke through the doors and dampers of a stove when an air flow reversal takes place in the chimney. It is caused by wind conditions or poor draft.

Banking the fire. The process of loading coal on the fire so that it will burn through the night.

Bituminous coal. A "soft" coal having considerable volatile matter which is given off as a smoky gas when heated. It is high in heat value and used mostly by the electric power industry.

Briquettes. Molded blocks or pellets of coal dust available in bag quantities.

Btu (British Thermal Unit). The amount of energy required to raise a pound of water 1° Fahrenheit. One thousand Btu will heat one gallon of water 120° F.

Caking coal. Coal that softens and adheres together at high temperatures.

Cannel coal. A noncaking coal that contains trapped oil. It burns with a bright, hot flame. It should not be used in closed stoves.

Carbon dioxide. A gas formed from the combination of the carbon in the coal and the oxygen in the air during the combustion process. It is nontoxic.

Carbon monoxide. A poisonous gas formed in a coal fire when not enough oxygen is present for complete combustion. Adequate draft must be present to exhaust this out the chimney.

Circulating stove. A radiating-type stove surrounded by a sheet-metal enclosure. The cool air is drawn from near the floor, heated by the stove, and moved through louvers near the top. A circulating pattern is set up within the room.

Clinker. Lump of melted ash. This forms when hot coals mix with the ash layer. Clinkers should be removed so that they don't block the grate.

Coke. Coal that has been heated to drive off the volatile matter. It is used as a domestic heat source in some areas of the United States. It burns similarly to hard coal.

Creosote. A tarry liquid substance deposited on the surface of the stovepipe or chimney flue lining when incomplete combustion of wood takes place and the flue gases cool to below 250° F.

Damper. A device used in pipes or ducts to control the flow of smoke or air.

Domestic water. The water used for cooking and washing, and in the bathroom.

Glossary

Draft. The slight vacuum or suction that occurs in the stove. The amount of draft differs with the type of equipment, the temperature of the flue gases, and the height of the chimney.

Draft regulator. A device placed in the stovepipe that can be adjusted to maintain a constant suction on the fire.

Excess air. Air admitted to a fireplace or stove that is in excess of that needed for combustion. Excess air should be kept to a minimum to limit the amount of heat carried up the chimney.

Factory-built chimney. A double-wall insulated or triple-wall approved chimney. It is often less expensive to install than a masonry chimney. It must be rated "*class A*" or "*all fuel*" to be used with a coal stove.

Fines. The dust and small pieces of coal that accumulate from mining and handling. They can be used in some large power plant furnaces and for briquettes.

Fireplace insert. A room-heating appliance that fits within the opening of a conventional fireplace. It is usually of double-wall construction so that additional heat can be gained from the fire and circulated within the room.

Flame retention burner. A new-type oil burner having higher burning efficiency due to greater mixing of the air and oil particles.

Flue gases. The products of combustion including smoke, carbon dioxide, moisture, and excess air that are vented through the stovepipe-chimney system.

Flue liner. A fireclay liner within the chimney that conducts the combustion gases to the outside. It is available in sections that must be cemented together with refractory mortar.

Fly ash. The small particles of ash that are carried up the chimney and escape to the outside.

Friability. The tendency for coal to crumble or break upon being handled. The lower-numbered friable coals create less dust.

Grate. A framework of iron bars that support the fire and allow combustion air to pass through. Generally cast iron is used.

Heat value. The amount of heat available as measured under ideal combustion conditions. Heat values allow fuels to be compared for their economic value. Heat values of coals vary from 7,000 to 15,500 Btu/per pound.

Ignition temperature. The lowest temperature at which combustion is self-sustaining.

Iron pyrite. A sulfide compound of iron found in coal. The sulfur is released when the coal is burned.

Lignite. A brownish-to-black coal, frequently woody in appearance. It is high in moisture content and breaks down to fine sizes when dried.

Listed. A stove model that has passed a comprehensive safety test by an accepted testing laboratory. Some states require that all new stoves sold be listed.

Make-up air. Air needed to replace that used in combustion. It normally enters the house through cracks around doors and windows. A separate supply duct can be used.

Multifuel furnace or boiler. A central heating unit that will burn two or more fuels. The most common units will burn wood or coal and oil or gas.

Peat. The product of partial decay of plants that accumulate in stagnant water in swamps and lakes. It is used as a fuel in some countries.

Primary air. Air for combustion that enters the stove below the grate. This air supply controls the intensity of the fire.

Radiant stove. A stove that has a single-wall firebox and warms the walls, furniture, and people in the room similar to the way the sun warms the earth.

Rank of coal. The classification system used to distinguish between the various types of coal.

Register. A heat transfer device that is placed in walls, floors, and ceilings to allow hot air to circulate within the house.

Secondary air. Air for combustion that enters the stove above the level of the fire. This air supplies the oxygen to help burn the volatile gases.

Slack. Dust, dirt, and small pieces left after coal has been screened.

Soot. A carbon substance that results from incomplete combustion and is deposited inside the stovepipe and chimney.

Spacers. Noncombustible brackets or stand-offs to hold the clearance reducer at least one inch away from a combustible surface.

Spark arrestor. Screen or other device built into factory-built chimney cap or installed on chimney to prevent sparks from landing on roof.

Stoker. An automatic coal feed device added to a central heating system.

Stove. A room-heating appliance which burns solid fuels and is free-standing. It can also be called a heater although this term is often reserved for larger heating devices.

Stoveboard. A prefabricated mat used for heat protection under a stove or on a wall.

Stovepipe. Single-wall metal pipe used to carry the flue gases from the stove to the chimney. It is not approved as an outside chimney material.

Subbituminous coal. A lower-rank coal high in moisture content and volatile matter.

Surface mining. When the coal seam is near the surface, the overburden soil and rock are removed, and the coal is mined using large power shovels and trucks.

Therm. 100,000 Btu of heat.

Thermostat. A temperature-regulating device used to control a heating or cooling system.

Thimble. A metal or clay connector that supports the stovepipe as it passes through a combustible wall. Adequate clearance around the thimble must be provided to prevent overheating of nearby combustibles.

Underground mining. The removal of coal from seams below the ground surface through shafts or sloped entries.

Volatile matter. The vapors, mostly hydrocarbons, that are driven off when coal is burned. The amount varies considerably with different ranks of coal.

APPENDIX:
STATE ENERGY OFFICES

Alabama Energy Management
  Board
Office of State Planning and
  Federal Programs
3734 Atlanta Highway
Montgomery, AL 36130

Division of Energy and Power
  Development of Commerce and
  Economic Development
MacKay Building
338 Denali Street
Anchorage, AK 99501

Office of Economic Planning and
  Development
Capitol Tower, Room 507
Phoenix, AZ 85007

Energy Conservation and Policy
  Office
3000 Kavanaugh Street
Little Rock, AR 72205

Conservation Division
1111 Howe Avenue, MS-60
Sacramento, CA 95825

Colorado State Energy
  Conservation Office
State Capitol, 1600 Downing
Denver, CO 80218

Under Secretary of Energy
80 Washington Street
Hartford, CT 06115

Delaware Energy Office
114 West Water Street
Dover, DE 19901

Energy Unit
Office of Planning and
  Development
Jackson School
Avon and R Streets, N.W.
Washington, DC 20007

Governor's Energy Office
301 Bryant Building
Tallahassee, FL 32301

Office of Energy Resources
270 Washington Street, S.W.
Atlanta, GA 30334

State Energy Office
Department of Planning and
  Economic Development
1164 Bishop Street, Suite 1515
Honolulu, HI 96813

Idaho Office of Energy
State House
Boise, ID 83720

Institute of Natural Resources
325 West Adams, Room 300
Springfield, IL 62706

Department of Commerce Energy
    Group
440 North Meridian Street
Indianapolis, IN 46204

Iowa Energy Policy Council
Capitol Complex
Des Moines, IA 50319

Kansas State Energy Office
241 West 6th Avenue, Room 101
Topeka, KS 66603

Kentucky Department of Energy
Bureau of Energy Management
Capital Plaza Tower
Frankfort, KY 40601

Research and Development
    Division
Department of Natural Resources
P.O. Box 44156
Baton Rouge, LA 70804

Office of Energy Resources
55 Capitol Street
Augusta, ME 04330

Maryland Energy Office
Suite 1302
301 West Preston Street
Baltimore, MD 21201

Executive Office of Energy
    Resources
73 Tremont Street, Room 701
Boston, MA 02108

Michigan Energy Administration
Michigan Department of
    Commerce
P.O. Box 30228
Lansing, MI 48910

Minnesota Energy Agency
980 American Center Building
150 East Kellogg Boulevard
St. Paul, MN 55101

Mississippi Office of Energy
Department of Natural Resources
P.O. Box 10586
Jackson, MS 39209

Missouri Energy Program
Department of Natural Resources
1518 South Ridge Drive
P.O. Box 176
Jefferson City, MO 65101

Conservation Bureau
Energy Division
Department of Natural Resources
    and Conservation
39 South Ewing
Helena, MT 59601

Nebraska State Energy Office
State Capitol Building
Lincoln, NE 68509

Nevada Department of Energy
1050 East William Street
Suite 405
Carson City, NV 89710

Governor's Council on Energy
2½ Beacon Street
Concord, NH 03301

New Jersey Department of Energy
101 Commerce Street
Newark, NJ 07102

Department of Energy and
    Minerals
P.O. Box 2770
Santa Fe, NM 87501

Appendix

New York State Energy Office
Agency Building No. 2
Rockefeller Plaza
Albany, NY 12223

Energy Division
North Carolina Department of
  Commerce
P.O. Box 25249
215 East Lane Street
Raleigh, NC 27611

North Dakota Office of Energy
  Management and Conservation
Capitol Place Office
1533 North 12th Street
Bismarck, ND 58501

Ohio Department of Energy
State Office Tower
30 East Broad Street
Columbus, OH 43215

Oklahoma Department of Energy
4400 North Lincoln Boulevard
Suite 251
Oklahoma City, OK 73105

Department of Energy
102 Labor and Industries Building
Salem, OR 98310

Governor's Energy Council
1625 North Front Street
Harrisburg, PA 17102

Governor's Energy Office
80 Dean Street
Providence, RI 02903

Governor's Office
Division of Energy Resources
SCN Center
1122 Lady Street
Columbia, SC 29201

Office of Energy Policy
Capitol Lake Plaza
Pierre, SD 57501

Tennessee Energy Authority
226 Capitol Boulevard Building
Suite 707
Nashville, TN 37219

Texas Energy and Natural
  Resources Advisory Council
411 West 13th Street
Austin, TX 78701

Utah Energy Office
231 East 400 South
Suite 101
Salt Lake City, UT 84111

Vermont State Energy Office
State Office Building
Montpelier, VT 05602

State Office of Emergency and
  Energy Services
310 Turner Road
Richmond, VA 23225

Washington State Energy Office
400 East Union
Olympia, WA 98504

Fuel and Energy Division
Governor's Office of Economic
  and Community Development
1262½ Greenbrier Street
Charleston, WV 25305

Department of Administration
Division of State Energy
1 West Wilson Street, Room 211
Madison, WI 53702

Energy Conservation Coordinator
320 West 25th Street
Capitol Hill Office Building
Cheyenne, WY 82002

ACKNOWLEDGMENTS

The information provided in this book is based in part on an extensive educational program developed as part of the Cooperative Extension Service programs at the University of Connecticut. The thousands of phone calls and letters that I have answered and the more than 100 slide presentations given throughout the Northeastern United States have given me the opportunity to assess the concerns and problems of many homeowners who have or will be installing a stove. Their concerns should also be yours.

I am indebted to the many stove shops, safety officials and coal burners that have shared their knowledge and expertise with me. I particularly wish to recognize Jim Mackey, Henry Kinne, and Marion Bradley whose resources and time I have used freely.

I am especially fortunate to have had Edward L. Palmer, my colleague, working with me these past several years. His helpful suggestions and critical review of the manuscript are very much appreciated.

Thoughtful remembrance is also made to my late grandfather John Bartok and the many other immigrants who labored in the coal mines for many years to start a beginning in their new homeland.

Finally I want to thank my wife, Janet, and children, Philip and Cynthia, who encouraged me to write this book.

CATALOG OF
MANUFACTURERS

Manufacturers and Importers

The following is a list of the addresses of manufacturers and importers of coal heaters, cookstoves, furnaces, boilers, and fireplace inserts described in this section.

Stove	*Company*
Allegheny	Allegheny 1001 Washington Blvd. Pittsburgh, PA 15206
Ashley	Ashley Heater Company 1604 17th Ave. S.W. Sheffield, AL 35660
Atlanta	Atlanta Stove Works 112 Krog St. Atlanta, GA 31706
Better'n Ben's	Hayes Equipment Corp. 150 New Britain Ave. Unionville, CT 06085
Buderus	R.W. Gorman Associates, Inc. 200 South Washington Ave. Washburn, WI 54891
Coal Stove Chubby	Plymouth Coal Stove Works 37 County Road (Rt. 106) Plympton, MA 02367
Combustioneer	Will-Burt Co. 169 S. Main St. Orrville, OH 44667
Crane	Crane Stove Works Box 2016 Duxbury, MA 02332
DeVille	Covinter, Inc. 70 Pine St. New York, NY 10005
Duo-Matic	Duo-Matic Olsen, Inc. 2510 Bond St. Park Forest, IL 60466
Enterprise	Enterprise Sales Route 2 Millersburg, OH 44654

Stove	Company
E-Z Insert	A-B Fab., Inc. P.O. Box 8867 Greensboro, NC 27410
Findlay Oval	Elmira Stove Works 22 Church St. Ontario, Canada N3B 1M3
Fireview	Comforter Stove Works Box 175 Lochmere, NH 03252
Franco Belge	Franco Belge Foundries of America 15 Columbus Circle New York, NY 10023
Hardin	Hardin Manufacturing Co. 3956 Highway 19 Longmont, CO 80501
Hitzer	Hitzer, Inc. 269 East Main St. Berne, IN 46711
Hoval (magazine feed boiler)	Arotek Corp. 1703 E. Main St. Torrington, CT 06790
HS Tarm	Tekton Corp. Conway, MA 01341
Hutch	Hutch Manufacturing Co. 200 Commerce Rd. Loudon, TN 37774
Jøtul	Kristia Associates 343 Forest Ave. P.O. Box 118 Portland, ME 04104
King	Martin Products P.O. Box 128 Florence, AL 35630
Koppe	Finest Stove Works P.O. Box 1733 Silver Spring, MD 20902
Lange	Svendborg Co., Inc. Box 5, Bridgeman Block Hanover, NH 03755
Leda	HDI Importers Schoolhouse Farm Etna, NH 03750
Liberty Bell	Liberty Bell Stove Works, Inc. 162 Reed Ave. West Hartford, CT 06110

Catalog of Manufacturers **163**

Stove	Company
LTD 4000	Jordahl Enterprises, Inc. 107 N. Box St. Kewanee, IL 61443
Mascot	Hiestand Distributors Route 1, Box 96 Marietta, PA 17547
Monarch	Malleable Iron Range Co. Beaver Dam, WI 53916
Olsberg	Svendborg Co., Inc. Box 5 Bridgeman Block Hanover, NH 03755
Pacific Princess	Pioneer Lamp & Stove Co. 105 S. Washington Seattle, WA 98104
Petit Godin	Bow & Arrow Imports 14 Arrow St. Cambridge, MA 02138
Pro-Former-Z	Pro-Former Engineering Corp. Mear Rd., Cochato Industrial Park Holbrook, MA 02343
Resolute	Vermont Castings, Inc. Prince St. Randolph, VT 05060
Riteway	Riteway Mfg. Co. P.O. Box 153 Harrisonburg, VA 22801
Shenandoah	Shenandoah Mfg. Co., Inc. P.O. Box 839 Harrisonburg, VA 22801
Stadler	Stadler Corp. P.O. Box 180 15 Lowell Rd. (Carlisle Ctr.) Carlisle, MA 01741
Surdiac	Classic Stove Works 233 Main St. Suite 608 New Britain, CT 00651
Timber-Eze	Timber-Eze, Inc. Route #5 Millersburg, OH 44654
Tirolia	Tirolia of America, Inc. 169 Dunning Rd. Middletown, NY 10940

Stove	Company
Titan	Kerr Controls Ltd. P.O. Box 1500 Truro, Nova Scotia B2N 5X2
Tritschler	Covinter, Inc. 70 Pine St. New York, NY 10005
Vigilant	Vermont Castings, Inc. Prince St. Randolph, VT 05060
Warm Morning	Locke Stove Company 114 West 11th St. Kansas City, MS 64105
Weso	Service Supply Corp. Div. of Texknit Pleasant Dr. Lochmere, NH 03252

Coal Heaters

The following is a listing of a variety of styles and types of coal heaters, furnaces, boilers, fireplace inserts and cookstoves. All dimensions, unless otherwise shown, are given as height by width by depth.

Manufacturer and Model	Type (radiant-R) (circulating-C)	Dimensions (Height by width by depth)	Weight (pounds)	Heating Capacity (Btu/hour)
ALLEGHENY				
Model A	R	28 × 30 × 21	400	4–5 rooms
Model C	R	32 × 24 × 23.5	400	4–5 rooms
ASHLEY HEATER CO.				
7150 C	C	36 × 35 × 21	275	4–5 rooms
BUDERUS-JUNO				
3608	C	24.8 × 18.6 × 12.7	150	12,800
3110	C	24.5 × 21.6 × 14	247	16,000
1012	C	27.3 × 22.9 × 13.5	227	20,000
COMFORTER STOVE WORKS				
Fireview	R	27 × 24.75 × 22	270	40,000–55,000

Catalog of Manufacturers

Manufacturer and Model	Type (radiant-r) (circulating-c)	Dimensions (Height by width by depth)	Weight (pounds)	Heating Capacity (Btu/hour)
CRANE STOVE WORKS				
Crane	R	27.5 (h) × 15(dia.)		40,000
DEVILLE				
7350	C	25.5 × 20 × 16.5	88	12,000–17,000
7771 T	C	28.5 × 34.25 × 15.75	215	30,000–40,000
FRANCO BELGE				
10-262	C	28 × 27.5 × 15.8	276	33,000
10-275	C	28 × 32.5 × 15.8	309	40,000
10-475	C	29.1 × 32.75 × 15	389	40,000
HAYS EQUIPMENT CORP.				
Better' n Ben's 801	R	30 × 21 × 15	310	40,000
HITZER				
55	R	35 × 17.5 × 26	260	50,000
75	R	40 × 21 × 30	310	80,000
HUTCH MANUFACTURING CO.				
Sunglow	C	28 × 22 × 16	240	800–1,000 sq. ft.
Cottage Stove	R	30 × 26 × 25	195	60,000
KING PRODUCTS				
9901-B	C	32 × 33 × 21	225	1,500 sq. ft.
King-O-Heat 330	R	32(h) × 16(dia.)	163	2 rooms
King-O-Heat 390	R	42(h) × 19(dia.)	271	6 rooms
KOPPE				
KK400	C	36 × 36 × 20	441	20,000
KK100/S	C	32 × 35 × 16	452	22,200
KK150/S	C	41 × 35 × 16	462	30,000
LANGE				
Ship's Stove 6732	R	22 × 13.5 × 13.5	165	3,000–5,000 cu. ft.
6304 RA	R	31.75 × 15.25 × 17	154	3,000–5,000 cu. ft.
LEDA				
Berliner K 6.14	C	31 × 25 × 13	375	22,400

Manufacturer and Model	Type (radiant-r) (circulating-c)	Dimensions (Height by width by depth)	Weight (pounds)	Heating Capacity (Btu/hour)
LIBERTY BELL				
8000 & 8002	R	22.25 × 18 × 18		50,000
LOCKE STOVE CO.				
Warm Morning 460	C	38 × 22.5 × 22	293	up to 5 rooms
Warm Morning 617-B	R	37 × 16.5 × 16.5	209	up to 5 rooms
OLSBERG				
Brilon 1.0	C	25.5 × 20.3 × 12.5	255	16,000
Herne 1.5	C	28 × 25.6 × 15.4	410	24,000
Marburg 1.5	C	31 × 29.7 × 15.7	440	24,000
PETIT GODIN				
3720	R	32 × 16 × 21	121	5,000–7,000 cu. ft.
3721	R	39 × 21 × 27	194	10,000–14,000 cu. ft.
PLYMOUTH STOVE WORKS				
Coal Stove Chubby	R	30(h) × 20(dia.)	220	60,000
RITEWAY				
37	R	41.5 × 26 × 34	415	73,000

Allegheny

Model C (coal/wood) Radiant. Contemporary, freestanding design. Double-wall heat chamber. Automatic thermostat and variable-speed rheostat controlling fan that delivers hot air up to 540 cubic feet per minute. Heavy steel-plate and cast-iron construction. Ash drawer, cast-iron shaker grate. Available in dark brown and flat black.

Allegheny Model C (left) and Model A (right).

Ashley

Model 7150 C (coal/wood) Circulating. Identical in appearance to the familiar Deluxe Imperial woodstove, but has internal differences that allow it to burn coal as well as wood. These differences: cast-iron rotating shaker grate, cast-iron retainers supporting firebrick walls. Special control valve on feed door to eliminate any pressure build-up which may occur in the combustion chamber while burning coal. Lift-up provides surface for emergency cooking. Large ash pan.

Catalog of Manufacturers **167**

Manufacturer and Model	Type (radiant-r) (circulating-c)	Dimensions (Height by width by depth)	Weight (pounds)	Heating Capacity (Btu/hour)
SHENANDOAH MANUFACTURING CO.				
R-76C	C	36 × 24 × 35.5	317	1,600–2,500 sq. ft.
R-76LC	R	33 × 18.5 × 32	244	1,600–2,500 sq. ft.
SURDIAC				
Petit Manor 508	C	28 × 28 × 18.5	213	38,000
Gotha 513	C	32.8 × 36 × 20	315	44,000
Royale 516	C	27.7 × 35.4 × 17.7	377	60,000
VERMONT CASTINGS, INC.				
Resolute	R	28.25 × 26.25 × 17	356	35,000
Vigilant	R	32 × 28.75 × 24.75	382	47,000
WESO	C	33.5 × 35 × 19	440	30,000
WILL-BURT CO.				
Combustioneer 35R	R	34 × 15.25 × 16.5	175	20,000–35,000
Combustioneer 65T	C	37.5 × 35 × 23	335	45,000–65,000

Ashley Model 7150C

Better'n Ben's

801 (coal/wood) Radiant. This is the same stove described under fireplace inserts but can also be a freestanding stove.

Buderus-Juno

Model 3608 (coal) Circulating. Large ash bin. "Bi-matic" control that automatically controls the burning of fuel. Brown enameled body, majolica-brown enameled front panel and top cover. Anthracite coal.

Model 3110 (coal) Circulating. German-made stove with sixteen models to choose from. Front panel glass window for viewing. Cast iron with majolica-brown enameled body, beige enameled front panel with grooved structure. Anthracite coal.

168 HEATING WITH COAL

Combustioneers (left to right): 65RM with cabinet package, 65RM, 65T, 35R, and 35R with cabinet.

Coal Stove Chubby

(coal) Radiant. Old-fashioned looking, barrel-shaped stove. Weighs 220 pounds. Top or front loading. Lower door is for ash cleanout. Single piece grate basket. Doors sealed with gaskets. Holds 35–40 pounds of anthracite coal.

Coal Stove Chubby

Combustioneer

Model 35R (coal/wood) Radiant, but has a cabinet to convert to circulating heater. Cabinet has lift-top for direct-heat cooking. Vertical box-like stove. Designed to burn primarily *bituminous coal*, although the model 35 and 65 heaters can burn anthracite coal. Cast-iron shaker grates, heavy twelve-gauge furnace steel construction, deep firebrick lining, manually controlled at two points.

Model 65T (coal/wood) Circulating. Cabinet model that can burn coal or wood without special adjustments or grate changes. Dial to set for your comfort level, automatic, nonelectric thermostat for damper control. Textured steel cabinet. Oversize combustion chamber. Split-lift top for cooking. Heavy 12-gauge (all models). Designed to burn primarily bituminous coal, but can burn anthracite. Model 65 RM converts easily with cabinet kit to gravity furnace and with a blower kit into a forced air system.

Comforter

Fireview Model (coal/wood) Radiant. Front door opens and lifts off to provide viewing of the fire. Spark screen, recessed fire chamber prevent ash or burning fuel from spilling out. Fireview performs as an efficient airtight heater when front door is closed. Both the Fireview and Standard models "interlock" seams are sealed with 3000° cement. Castings are twice the thickness of regular stove plate for greater durability and extra thermal mass. Preheating chamber for combustion efficiency. Small hot spot cooking/kettle area. Cast-iron construction. Other models available, anthracite nut or stove coal.

Crane

Model 44 (coal) Radiant. Vertical box style stove. Double baffled, airtight. Large shaker grate system. Glass view. Heavy, high temperature firebox lining. Anthracite coal.

Crane coal cooker/heater

DeVille

Model 7771T (coal/wood) Circulating. DeVille is France's largest manufacturer of heating stoves. Firebrick lined chamber. Hopper with gravity feed lengthens burning time. When hopper is removed this model becomes a woodburning stove that can take up to 23-inch logs. Porcelain enamel finish coat. Pyrex-glass window for fire viewing. Other coal models to choose from. Anthracite coal.

Franco Belge

(coal) Circulating. Four models available. Decorator cabinet styling. Heavy-gauge steel cabinet with porcelainized enamel finish. Eight-position automatic thermostat. Hot plate for warming food located under hinged top cover. Adjustable hopper. Large heat-resistant glass door gives clear view of fire. Top-loading. Cast-iron construction.

Hitzer

75 (wood/coal) Radiant. Designed and manufactured by Swiss Amish people of Berne, Indiana. For heating water, 22-gallon galvanized jacket and a 1-gallon water cap are available. Other models to choose from. Automatic thermostat.

75-C (coal/wood) Circulating. Cabinet model of model 75.

55 (coal/wood) Radiant. Smaller version of the 75-C. Cabinet available for this model also.

Hutch

Sunglow (coal/wood/charcoal) Circulating. Cast iron with an enamel finish. Modern appearance. Large firebox and lift top for convenient loading. Thermostatic draft control. Unique internal circulation systems designed to improve combustion of fuel and hot gases. Built-in ash drawer, grate shaker handle.

Cottage Stove (coal/wood) Radiant. Ideal for cottages, enclosed porches, recreation rooms, etc. Old-fashioned design with black finish and brass top rail. Top door swings down for easy fueling. Ash pit door lifts off for easy cleaning. Two 8-inch cooking plates. Complete with fittings for wood or coal operation. Cast iron.

King

Model 9901-B. (coal/wood) Circulating. Cabinet model with lifetime porcelain finish. Automatic thermostat. Extra large ash drawer. Heavy duty firebox. Full end cabinet door for complete access to ash and loading doors. Combustion air vent prevents smoke and flame flashback when feed door is opened. Egg-sized anthracite or bituminous.

King-O-Heat Models (coal) Radiant. Vertical style heater with four models to choose from. Egg-sized anthracite or bituminous.

Koppe

Model KK 400 (coal/wood) Circulating. German-made, contemporary ceramic tile stove. Viewing window. Automatic control, heat sensor opens and closes draft as needed for uniform output. Fireclay lined, cast-iron firebox. Heat by convection and radiation. Large ash pan. Anthracite, good quality low volatile nonfusing bituminous coal, pressed coal briquettes. Nut or stove coal. If not available use egg coal.

Model KK 150/S (coal/wood) Circulating. Ceramic tile heater with natural stone top. Nine other models to choose from with tiles available in special patterns and colors. Anthracite, good quality low-volatile nonfusing bituminous coal. Nut coal. If not available use egg coal.

Lange Model 6304RA

Lange

Model 6304 RA (coal) Radiant. Manufactured in Denmark. Vertical design for small cabin or den. Cast iron, firebrick lined, ringed lids for cooking. Available in red, green or black enamel. Pea or nut coal.

Ship's Stove 6732 (coal) Radiant. Compact ship's stove styled like a seafarer's trunk. Cast iron, firebrick lined, bolts firmly for safe installation in boats. Removable rings in cookplate permit pot to sit securely, brass protective rail encircles cooking surface. Pea or nut coal.

Leda

Berliner Model K6.14 (coal) Circulating. Manufactured in West Germany. Ceramic tile stove, cast-iron firebox is lined with firebrick. Top and floor plate are porcelainized enamel over cast iron. Baffled at the top inside to retard heat travel. Doors on each model are provided with permanent gasket. Cook-

Koppe Model KK 150/S

ing surface under the top lid. Berliner *Model 6.14 R* has automatic regulator. Ashes are shaken into the ash drawer by means of a handle outside the stove that rotates the grates inside. Anthracite coal.

Liberty Bell

(coal/wood) Radiant. ¼-inch steel firebox, 14-gauge ashpan. Cast-iron doors, door frames, dampers, grates, and inside front frame. Firebrick. *Model 8002*—Free-standing. *Model 8000*—Fireplace. Anthracite coal.

Liberty Bell Model 8000

Olsberg

Model Marburg 1.5 (coal) Circulating. Made in Germany. Internal stove cast iron. Top, front door, sides and floor plate are majolica brown and enameled. Top has slits for warm air circulation. Glazed tiles are covered with gold-brown varnish. Glass door, complete cast iron with outside porcelainized finish, tile sides, thermostatic control. Combination regulator enabling automatic or hand-directed slow burning. Five other models available. Pea or nut coal.

Petit Godin

(coal/wood) Radiant. Design of stove little changed since 1889. Aesthetically interesting, compact French cylindrically shaped heater. Two attractive models in four colors. One hundred percent pig iron with fine porcelain enamel applied at 1700° Fahrenheit. Warming surface hidden under round top of stove. Firebrick lined. Secondary air channel. Nut-sized anthracite coal.

Riteway

Model 37 (coal/wood) Radiant. Riteheat regulator requires no electricity. For even heat set regulator at desired temperature. Cast-iron flue baffle, direct draft damper. Stainless-steel water heater available for this model for heating domestic water. Riteway cabinets available for Model 37, adapting this model to a circulating heater. Fireplace accessory installed allows heater to supplement hydronic heating system up to 11,000 Btu/hour while it is being used as a radiant heater.

Riteway Model 37

Shenandoah

Model R-78LC (coal/wood) Radiant. Airtight design, bimetal thermostat control, heavy gauge steel construction, firebrick-lined firebox, cast-iron shaker grate, large ash pan. Coal draft baffle. *Model R-78C* is a coal heater with porcelain cabinet. Optional blower for cabinet models.

Surdiac

Surdiac Royale 516 (coal) Circulating. Cast-iron firebox. Modern design with enameled steel body. Imported from Belgium. Thermostat, easy visibility of fire producing strong radiating heat, backed with large heat exchanger for convection heat. Available in several other models and several colors. Pea-sized anthracite coal.

Surdiac Royale 516

Vigilant

Vigilant (coal) Radiant. Design inspired by its larger brother, the *Defiant*, and by architectural heritage of America near end of eighteenth century. Available in both wood-burning and coal-burning models. Identical to its wood-burning brother on outside, the coal-burning *Vigilant* features "stove within a stove" design which employs gravity-feed fuel magazine for automatic maintenance of fuel bed depth. Coal stove's front doors feature high-temperature glass panels for view of fire. Large polished cooking griddle. Two folding drying racks that tuck out of way when not in use. Coal stove can easily be converted to a woodburner by the removal of coal-burning components. Premium grade nut- and pea-sized anthracite coal.

Warm Morning

Model 617-B (coal) Radiant. Natural grey steel bodies, cast-iron tops, feed and ash door sections and bases finished in baked black enamel. All have side-hinged feed doors and reversible flue pipe collar. All models have patented four-flue firebrick construction. Other models available.

Model 460 (coal) Circulating. Porcelain enamel cabinet with louvers for improved appearance and performance. Front and side vents give improved radiation and heat circulation. Large front feed door. Another larger model available, heats up to six rooms, built-in automatic thermostat, cabinet side doors may be opened for quick radiant heat release. Also two-speed electric fan available for larger model.

Warm Morning Model 460

Weso

(coal) Circulating. Ceramic tile stove imported from Germany. Modern version of the magnificent tile and masonry Kachelofens that have been warming European homes for over 700 years. Double cast-iron firechamber, surrounded by wall of hand-glazed terra cotta tiles with an air space in between. Removable tile sections. Automatic thermostat. Enclosed ash collector pan that removes easily. Utilizes pre-heated primary and secondary air for maximum combustion efficiency. Kettles simmer gently on top grille. Raise grille to cook directly on cast-iron surface. Top and front grilles are black-enameled cast iron in wrought-iron motif. Chesnut or egg anthracite.

Weso parlor stove

Coal Boilers

Manufacturers and Models	Dimensions (inches) (h × w × d)	Water Capacity (gallons)	Weight (pounds)	Heating Capacity (Btu/hour)
BUDERUS				
Logana 02.40-14	38.75 × 16.5 × 15.3	6.25	450	56,000
Logana 02.40-27	38.75 × 16.5 × 23	8.6	622	108,000
Logana 02.40-40	38.75 × 16.5 × 35	12.1	877	160,000
FRANCO BELGE				
Forestiere 93.27	50 × 28 × 23.5		748	60,000–108,000
Forestiere 93.40	52 × 28 × 30		836	100,000–160,000
HARDIN				
FCA (PK) 450	64 × 22 × 21.6		1450	75,000–100,000
FCA (PK) 790	64 × 22 × 34.6		2200	200,000–300,000
HS TARM				
OT-35S	51 × 39.5 × 30	76	1089	112,000
OT-70S	51 × 46.75 × 39.5	130	1800	196,000
MB-Solo 30	49 × 17.75 × 25.25	29	594	72,000
MB-Solo 75	49 × 21 × 41.5	55	1177	180,000

Manufacturers and Models	Water Dimensions (inches) (h × w × d)	Capacity (gallons)	Heating Weight (pounds)	Capacity (Btu/hour)
MASCOT				
MC200	45 × 24 × 28		900	110,000
RITEWAY				
Riteway RB 75	59 × 33.5 × 42		1400	75,000
STADLER				
KL 21	41.9 × 36.5 × 22.1	42	704	68,000
H 33	41.6 × 24.8 × 31.1	25	726	100,000–132,000
UL 90	47.6 × 28.3 × 38	51.5	1342	288,000
TITAN	56 × 26 × 31			
TRITSCHLER				
HK 2516	42.5 × 17.1 × 18.75	9	395	62,400
HK 3825	42.5 × 19 × 22.6	15.75	483	97,500
HK 6040	42.5 × 24.5 × 27	30.5	693	155,000

Buderus

Logana 02.40 (coal/wood) A cast-iron sectional boiler suitable for solid fuel firing. Conversion to oil or gas is possible. Flue box with damper and damper positioner, loading door, ash door with damper, deep ash pan, threaded flanges for boiler flow and return, cleaning brushes, firing controller, operating and assembly instructions. Blue boiler jacket and heat insulation. Five models available.

Franco Belge

Forestière Model 93.40 (wood/coal) Central heating boiler burns wood/coal and is easily adapted to burn oil. Large hearth. Rate of burning thermostatically controlled from temperature of the central heating water. Design of heat exchanger with narrow air passages improves combustion and heat transfer. Heat exchanger made of thick steel. Large ash pan. Another model available (93.27).

Hardin

FCA (PK)450 and FCA (PK)790 (coal) Each boiler in this series is a horizontal tubular unit. Outer steel jacket with high-density insulation insures minimum standby heat loss. Tankless coils (extra) are located in the boiler to provide full-rate capacity for domestic hot water. Accessories include: Water—combination thermostat, altitude and pressure gauge, relief valve. Steam—steam gauge, water site glass gauge, and safety pop valve.

HS Tarm

OT Models (wood/coal, gas/electricity) Steel plate boiler for modern residential hydronic heating and domestic hot water supply. Manufactured in Denmark. Wood can be used in combination with oil, gas, or electricity (with installation of optional electrical heating elements). On the left side combustion chamber for oil or gas, on right, large firebox for wood. With the addition of shaker grates, coal or coke can be burned. Tankless system for the production of domestic hot water. Can supply ample hot water for large homes with several bathrooms.

Tankless coil is ¾ inch copper tubing and can convert entire output of boiler to hot domestic water. Automatic draft regulator for wood.

Catalog of Manufacturers   175

HS Tarm type OT multifuel boiler

Cast-iron grates. Cleaning tools. Built-in, glass-lined tank for domestic hot water. Insulated jacket with orange-red baked enamel finish. Pressure relief valve for hot water coil. High limit (overheat) control. All steel plate exposed to flue gases is over ¼ inch thick. Water capacity 73.5 to 130 gallons on OT models. Maximum hot water output 2.3 to 5.7 gallons per minute. Several models. Nut-sized anthracite coal.

Mascot

MC200 (coal/wood) ¼ " boiler plate steel throughout with heavy cast-iron doors, frames, and grate assembly. Dual damper control system, large sealed ashpit, flip-over grates, 12" × 12" access door. Firebrick-lined firebox. Honeywell temperature and safety controls.

Riteway

Riteway RB-75 (coal/wood, oil/gas) Two separate combustion chambers—one for coal/wood and one for oil/gas. Built-in skid, allowing easy moving and installation of unit. Hinged ash door, large fuel door for easy loading of wood, extra large solid fuel combustion chamber. Equipped with combination temperature, pressure and altitude gauge, pressure relief valve, barometric draft regulator, and complete thermostatic controls. Optional painted steel jacket useful for installations requiring minimal heating of boiler room.

Stadler

KL-Series (wood/coal, oil/gas). Two separate combustion chambers, flue passes, and smoke pipes. Automatic switchover to oil or gas operation should wood or coal fire go out. Two primary water connections to accommodate all piping situations. Large temperature-controlled domestic hot water tank. Porcelain enamel tank lining and magnesium anode provide protection against corrosion. With or without temperature-controlled domestic hot water tank, designed for central hot water heat systems. Four models available, ranging in output from 68,000 to 144,000 Btu/hour.

UL-Series. (wood/coal, oil/gas) With or without temperature-controlled domestic hot water tank, designed for central hot water heating systems. Can easily

Mascot MC200

be changed over to solid fuel firing by means of an optional conversion kit. Complete water jacket. Built-in preheating baffle for the primary return water prevents corrosion damages caused by internal condensation. Large cleanout doors. Provisions for installing either an external tankless water heater or a Stadler temperature-controlled domestic hot water tank are standard. Porcelain enamel tank lining and magnesium anode provide maximum protection against corrosion on all Stadler domestic hot water tanks. Prewired electrical control system, with three aquastat functions, two thermometers, and burner control. Seven models available, ranging in output from 80,000 to 288,000 Btu/hour.

H-Series (coal/wood) Installed as solid fuel-fired unit in combination with any existing oil- or gas-fired boiler, will give owner all advantages of multi-fuel control heating system with automatic switchover to oil or gas operation should wood or coal fire go out. Installed as solid fuel-fired primary heating unit, boiler will carry the entire heating load for heating system, as well as domestic hot water, without an oil or gas back-up. Domestic hot water supply is assured through the Stadler H-series boiler by means of the existing hot water system or an externally connected hot water unit. Large top filling door makes access to the firebox convenient. Built-in baffle for the primary return water prevents corrosion damage caused by internal condensation. Large accessible cleanout doors. Flue collar can be installed on either side or back. Four models available, ranging in output from 100,000 to 340,000 Btu/hour.

Titan

Titan (coal) Large combustion chamber lined with firebrick provides controlled burning. Preheated air is introduced over the fire, igniting burnable gases. Zone valves and circulator are automatically controlled. Will perform as a

Kerr Titan coal-fueled boiler

gravity furnace in the event of a power failure. Should boiler overheat, the thermostat is automatically overridden and heat is directed into the living area. An additional safety system reduces the boiler temperature by releasing steam directly over the fire. Rugged cast-iron doors and quarter-inch plate in the heat exchanger. Jacket is rust-resistant, satin-coated (urethane enamel). Bituminous and anthracite coal.

Tritschler

HK Series (coal/wood) German-made coal/wood central heating boiler. Range of 62,000 to 156,000 Btu/hour. Can be converted to gas or oil. Can be hooked up to supplement an existing oil boiler. Six-inch flue. Efficiency of boiler improved with a turbulator. Equipped with safety heat sink which will absorb heat above 212° Fahrenheit. Water capacity 9 to 30.5 gallons. Five models available.

Coal Furnaces

Manufacturer and Model	Dimensions (inches) (h × w × d)	Weight (pounds)	Heating Capacity (Btu/hour)
DUO-MATIC/OLSEN, INC.			
CWO-B 112	51.25 × 44.1 × 48.25	970	112,000–123,000
CWF	52.6 × 27.6 × 33.75	625	120,000
CWO-B 140	51.25 × 44.1 × 48.25	970	134,000–151,000
JORDAHL ENTERPRISES			
LTD 4000	47 × 24 × 33.25	750	90,000–120,000
RITEWAY			
LF-20	52 × 32 × 50	1186	125,000
SHENANDOAH			
Shenandoah F-77C	44.5 × 24 × 45.5	423	75,000
Shenandoah AF-77C	44.5 × 24 × 32	348	75,000
TIMBER-EZE			
104-RE	34 × 24 × 34	750	120,000
WILL-BURT COMPANY			
Combustioneer 24 FA (with stoker S20)	57.25 × 61.4 × 42.25	725	156,000

Combustioneer

24 FA (coal) Furnace body is heavy boiler-plate steel, precision welded. Smoke and gas tight. Firebrick lining made to withstand heat up to 3050° Fahrenheit. Smoke consumer creates better combustion. Large boiler-plate steel, doughnut-type radiator is welded in one piece. Heavy baffle causes the hot gases to circulate more effectively. Nine-inch flue pipe. *Stoker model # S20.* Stoker and furnace must be ordered separately.

Duo-Matic

CWO-B Series (wood/coal/gas/oil) One side burns coal/wood, the other burns gas/oil. Dual combustion chambers, dual thermostatic controls. Solid fuel side lined with 2½-inch heavy-duty firebrick rated for up to 3000° F. Cast-iron grates for either wood or coal. Easy access to convenient removable ash pan. Blower section assembles on right or left side of heat exchanger. Uses regular size spun glass disposable filters. One-piece airtight doors. Seven-inch flue pipe.

CWF Model (coal/wood) Add to present warm air furnace or install as free-standing solid fuel furnace. Heavy-duty welded construction. Secondary heat exchanger to extract the maximum amount of heat. Heavy-duty cast-iron grates for either coal or wood. Lined with firebrick rated for up to 3000° F. Uses regular size spun glass dispos-

Duo-Matic Olsen furnace

able filters. Optional blower section assembles on right or left hand side of heat exchanger. Ash pit door has easy access to convenient removable ash pan. Automatic thermostat controls. Seven-inch flue pipe.

Jordahl

LTD 4000. (coal/wood) Heavy duty 14 x 14-inch cast-iron doors with adjustable draft flow. 2½-inch firebrick, cast-iron grates, ash pan. Two Dayton 465 CSM blowers. Honeywell limit control switch. Eight-inch flue. Designed to be used as supplemental or independent heating system. It uses the air ducts and chimney of your present gas or oil furnace. The forced air from the LTD 4000 activates the blower of your present furnace and circulates the warm air through your present furnace heat ducts. When the temperature between the interlining of the stove reaches 150° Fahrenheit, blower will engage and continue to transfer the heat produced by the wood-coal burner until the interjacket temperature goes below 120° Fahrenheit. In the summer, the

Jordahl LTD 4000 furnace

LTD 4000 can be used to transfer cool air from the basement to your living quarters.

Riteway

Riteway LF-20 (wood/coal) Wall thermostat, barometric damper, forced-draft and warm air circulation blowers, filters, heat exchanger, belt-drive blowers. Domestic water heater available rated at 3.5 gallons per hour designed to be used with storage tank or existing water heater. Jacket available.

Riteway LF-20

Shenandoah

Shenandoah F-77 (wood/coal) Wood furnace designed to easily convert to coal. Heavy gauge aluminized steel firebox for extended life. Seams welded for airtight construction. Satin black painted steel cabinet. Permanent aluminum filter. Removable panel provides easy access to the filter and the 900 CFM blower unit at the rear of furnace. Equipped with automatic blower control for maintaining plenum temperatures. Manual blower switch for summer use. Can be installed on either side of an existing furnace. The primary furnace is then used only as additional heat is required. Bimetal thermostat to control rate of burning and heat output. Primary and secondary air intake. Firebrick lining. Large access door. Ash pan, cast-iron shaker grate, flue inspection port, smoke baffle.

Shenandoah F-77C. Wood model converted to use with coal with *76-CK Kit*: cast-iron outer grate to replace sheet-steel grate, and coal draft baffle to replace wood draft baffle.

AF-77 Series. (wood/coal add-on furnaces) Designed for use with existing forced air heating systems. Heavy-gauge steel firebox. Cool air return can be on either side or rear of cabinet; automatic blower control, manual blower switch for summer use. Bimetal thermostat to control rate of burning and heat output. Primary and secondary air intake. Firebrick lining, ash pan, cast-iron shaker grate, smoke baffle. Easy conversion from wood to coal.

Timber-Eze

Model 104-RE (coal/wood) Designed as a coal-burning companion to an electric, oil, gas, or solar hydronic heating system. Heavy cast-iron rocking grates, a firebox lined with firebrick, large firebox with large opening. Automatic hydronic actuated draft control. Features 3/16-inch and ¼-inch plate steel construction, heavy-duty cast-iron grates and decorated cast-iron doors. The *Timber-Eze* is available in five models: two hydronic add-on boilers, two radiant stoves, and a hot air add-on furnace.

Fireplace Inserts

Manufacturer and Model	Dimensions (inches) (h × w × d)	Weight (pounds)	Heating Capacity (Btu/hour)
ASHLEY	24 × 32.75 × 18	330	1,500– 2,000 sq. ft.
HAYES EQUIPMENT CORP. Better' n Ben's 801	30 × 21 × 15	310	40,000
PRO-FORMER-Z			
Z-24	25 × 24 × 23.5		14,000 cu. ft.
Z-28	25 × 28 × 23.5		20,000 cu. ft.
Z-30	25 × 30 × 23.5		24,000 cu. ft.
SHENANDOAH FP-1	25 × 26 × 20	400	35,000

Ashley

(coal/wood) Fits inside most existing fireplaces. With doors closed, the Ashley fireplace insert becomes a heat circulator. Manually adjustable draft regulators and airtight seal. With doors open it becomes a fireplace. Optional cast-iron coal grate easily converts this to a coal-burning unit. Smoke outlets at 45° allow installation in shallow fireplaces. Cast-iron, gasketed doors provide airtight seal and fire retention. Heat output greatly increased with addition of an optional blower that forces air around the firebox. Blower is thermostatically controlled (manual and automatic). Surround panel is 18-gauge steel backed with insulation for exact fit.

Ashley fireplace insert

Better' n Ben's Model 801 FP fireplace stove

Better'n Ben's

Model 801 FP Fireplace Stove (coal/wood) Cast-iron doors, non-asbestos door gaskets, "cool touch" metal handles, fully firebrick lined, external shaker grate system, ash pan, shaker rod firetender, secondary air system, rear convection shroud, large safety glass window, cast-iron baffle system. Optional 135 c.f.m. blower system can be attached to the baffled convection shroud. *Model 801 FS* is freestanding model. Nut-size anthracite coal recommended.

Pro-Former-Z

(coal/wood) Three sizes available. Viewing-type unit has large enclosed Pyrex glass front area. Can be used as airtight unit. Blower system has the capacity to circulate all air in an average home over the hot surfaces of heating unit every 30 minutes. Shaker grate, removable ash pan. Blower attached to rear base of unit. Anthracite coal.

Shenandoah

Model FP-1 (coal/wood) Designed to fit any fireplace ranging in width from 30 to 40-inches and in height from 25 to 30 inches. Depth of unit in the fireplace is variable from fully inserted to extending out on the hearth 8 inches with a minimum fireplace depth of 14 inches. Designed to provide a fireplace atmosphere with the efficiency of airtight

Shenandoah Model FP-1

construction. Has bimetal thermostat draft control, glass doors, gasket-sealed doors for efficient operation, firebrick lining, removable variable-speed twin blowers, manual-automatic blower switch. Thermostat control to allow blowers to operate only when sufficient heat exists in firebox. Double-wall construction for circulating air. Ash pan, aluminized steel firebox, cast-iron grate, heavy steel construction, shaker grate, fire screen. Legs for free-standing applications *(FP-S)*. *Model FP-S* has top cooking surface.

Coal Cookstoves

Atlanta

Model 15-36—Style 3 Old-fashioned design with cast-iron cooking surface and oven. Warming ovens at top. Cooking surface contains four larger lids and two smaller lids.
Dimensions: 29.5 × 35.5 × 21.5 inches
Oven: 15 × 14 × 11 inches
285 pounds

DeVille

Model 8545 Cooker/Boiler (wood/coal) Contemporary design, top made of polished cast iron, oven has self-cleaning walls and lighted oven window. Fire

controlled through an eight-position thermostat and extra damper. Double shaker grate, large ash pan for easy maintenance. Storage space for cooking utensils and pans in base. Four to eight radiators plus one water tank can be hooked to this stove.
54,000–74,800 Btu/hour
33 × 34 × 24 inches
Oven: 11 × 18 × 17 inches
397 pounds

Model C 46 Cooker/Boiler (wood/coal) Heat-resistant chocolate brown porcelain enamel cooker/boiler with a wide polished cast-iron cooking surface.

Large continuous-cleaning oven with see-through door. Storage drawer underneath oven. Two shaker grates. Coal loaded from the top or from front. Optional thermostat for controlling the oven temperature, oven light switch. Oven temperature from 350° to 625° Fahrenheit. 24,000 Btu/hour (radiant heat); 72,000 Btu/hour (central heat). 32.8 × 33.46 × 25.63 inches
Oven: 11 × 14.75 × 16.5 inches
620 pounds

Enterprise

Model 52 DE Savoy (wood/coal) Contemporary design, woodgrain facing on background, fluorescent light, clock with minute-minders. Porcelain finish. Cast-iron cooking surface with two lids (one two-section). Available in white, black, or almond enamel.
32.5 × 31.5 × 23 inches
Oven: 16 × 19 × 11.5 inches
365 pounds

Model 52 RHS Savoy (wood/coal) Contemporary extended model range provides extra working surface on top and storage space inside. Available in black or white enamel.
32.5 × 39.5 × 23 inches
Oven: 16 × 19 × 11.5 inches
6.75 imperial gallon tank heats hot water.
415 pounds

Oval

Model 9919 3W (coal/wood) Old-fashioned design. Uses gasketed doors and bell dampers to control primary air supply. Unique wood grate easily adjustable for summer. Cooking surface contains five 9.5-inch lids and one three-section lid. Solid-copper water reservoir, 6.5 imperial gallon capacity. Large warming closet (approximately 140° Fahrenheit).
1,500 square feet radiant heat.
32 inches (height to cooking surface), 58 inches
Cooking surface: 24 (h) × 34 (d) inches

Franco Belge

In both stoves listed below top surface made of polished cast iron. Enameled oven evenly heated from five sides. Convenient warming drawer. Adjustable air inlet controls the burning rate and heat output.

Model 44-147 20,000 Btu/hour.
32.3 × 29.5 × 23.2 inches
Oven: 12.5 × 12.5 × 15.5 inches
358 pounds

Model 64-148 25,000 Btu/hour.
32.3 × 33.5 × 23.2 inches
Oven: 12 × 16 × 15.5 inches
Back boiler for domestic water
403 pounds

Model 1707 Cooker/Boiler (coal/wood) Heats up to fifteen radiators. Huge "whole-top" hot plate. Main oven for roasting and baking at temperatures up to 450° Fahrenheit. Second oven for slow cooking and warming at temperatures up to 260° Fahrenheit. Summer grates available.
67,000 Btu/hour
31.5 × 39.4 × 31.9 inches
Main oven: 9.6 × 14.2 × 21.7 inches
Slow oven: 6 × 13 × 14 inches

Jøtul

Model 404 (coal/wood) Compact Norwegian design takes minimum space. Heavy cast-iron construction. Adjustable damper directs heat to the oven or cooking plates for boiling or

Oval cookstove

baking. Front grating permits the stove to be fired with either coke or pea-sized coal. Two lids plus one two-section lid on cooking surface.
31.5 × 24.5 × 20 inches
223 pounds

Monarch

Model R9CW (wood/coal) Contemporary design. Two polished iron 8" lids and anchor plates, unbreakable, heat evenly. Large firebox, large ash pan with non-spill handle. Right side of range has a stainless steel reservoir (4 gallons) with built-in humidifier as standard equipment. Exterior finish of titanium porcelain enamel. Large warmer/storage compartment.
34.6 × 43 × 25 inches
Oven: 19 × 15.75 × 21 inches
513 pounds

Monarch Model R9CW wood-coal range

CE119Y-1 Duo-Oven (coal/wood/electric range) Electric cook top equipped with 8" frying and three 6" surface units. Large firebox with duplex grates. Large ash pan. King-size "Duo-Oven": 19-inch oven heated by coal, wood, or electricity, individually or in combination. Coal/wood heat may be supplemented by electric heat to keep oven at desired temperature. Automatic oven with timer. Electric cooking section is protected from coal/wood heat.
35 inches height to cooking surface.
47.74 (overall height) × 43 × 25 inches
537 pounds

Catalog of Manufacturers **183**

Model 6LEH (coal/wood) Electric range with built-in wood or coal kitchen heater.
47.5 × 36 × 26.1 inches
Oven: 20 × 20 × 17 inches
327 pounds

Pacific Princess

All iron. Old-fashioned design. Warming closets. Five-gallon stainless steel reservoir. Victorian fold-down trivets to hold pot off fire. Improved temperature gauge. Polished griddle.
30 (height to cooking surface) × 40 × 28 inches
57 inches overall height
325 pounds

Tirolia

7N (coal/wood) Grates adjustable for summer and winter use. Enameled stove lid. Manually controlled burning rate. Extra large oven thermometer. Brick-lined firebox.
34.8 × 35.5 × 23.6 inches
Oven: 13.8 × 15.7 × 19.1 inches
547 pounds

7HT (coal/wood) Grates adjustable for summer and winter use. Enameled stove lid. Manual controlled burning rate. Extra large oven thermometer. Steel hot water jacket (for domestic water).

Monarch Model 6LEH (36-inch electric range with built-in kitchen heater)

34.8 × 35.5 × 23.6 inches
Oven: 13.8 × 15.7 × 19.1 inches
1.45 gallons—approximate volume of boiler
560 pounds

7ZH (coal/wood) Grates adjustable for summer and winter use. Enameled stove lid. Automatically controlled burning rate. Firebox lined with ¼" steel hot water jacket.
34.8 × 35.5 × 23.6 inches
Oven: 13.8 × 15.7 × 19.1 inches
635 pounds

These are Austrian stoves manufactured since 1919. All contemporary design, iron and steel construction. Available in white or harvest gold.

INDEX

Accessories, 63 and *illus.*, 64
  poker ("fiddle stick"), 107
Advertising, inaccurate, 64–66
Air circulation, 39, 43, 53, 63 and *illus.*, 64, 95
  for furnace, 127
Air leaks in stove, 53
Air quality, 11–13
Air supply:
  for furnace, 126, 138 and *chart*, 139
  for stove, 36, 47–48, 95–97 and *illus.*, 101
Amount of coal needed, 4 and *table* (#1), 29–31
Anthracite coal, 19
  burning procedures, 104–109 and *illus.*
  sizes, *table* (#4) 23
Ash:
  content, of coal, 24–25 and *illus.*
  elements in, *table* (#10), 116
  and garden, 13
  removal, 10, 50–51, 116
  shaking, 107, 116; *see also* Grate
Automatic coal burner, 135–38 and *illus.*

Bacharach test, 54
Banking fire, 107–108, and *illus.*
Bin (for storage), 32–33 and *illus.*
Bituminous coal, 18
  burning procedures, 110–113 and *illus.*
  sizes, *table* (#4) 23
Blower, 39, 43, 53, 63 and *illus.*, 64, 95
  for furnace, 127
Boiler:
  definition, 123
  design, 130–32 and *illus.*
  draft requirements, 138 and *chart*, 139
  hand firing, 139
  installation, 132–34

Boiler installation *(cont'd)*
  of add-on unit, 133–34 and *illus.*
  maintenance, 139–40
  manufacturers (list), 173–76 and *illus.*
  multifuel unit, 134–35
Breaking in stove, 103–104
Briquettes, 28
Burning process, 34–36; *see also* Firing stove
Buying coal, 3–4, 27–31
  cooperative, 28–29
Cannel coal, 18–19
  for fireplace, 115
Cast iron stove, 44–45
Ceiling:
  installation of chimney in, 91–93 and *illus.*
  insulation of, 7, *map*, 8
Central heat, 123–41 and *illus.*
Charcoal, 103
Chimney:
  ceiling installation, 91–93 and *illus.*
  cost, 6–7
  fire, 117
  fireplace, 84–89 and *illus.*
  for furnace, 126–27, 138 and *illus.*
  inspection, 82–83 and *illus.*
  installation and adaptation, 79–84 and *illus.*, 89–95 and *illus.*
  new, 89–95 and *illus.*
    factory built, 90–91
    masonry, 90
  wall installation, 93–95 and *illus.*
Circulating stove, 42 and *illus.*, 43
Clearance from combustibles, 72–76 and *illus.*, *table* (#8) 73
  of furnace, 128–29 and *illus.*
  of stovepipe, 77 and *illus.*, *table* (#9) 78
  from windows, 75
  zero-clearance fireplace unit, 98–99

185

Clinker, 107
Coal:
  amount needed, 4 and *table* (#1), 29–31
  cost of, 4, 6, 27–29
  delivery, 27–28
  formation of, 15–19
  as a fuel, 15–36
  ignition temperature, 4, 35 and *table* (#6)
  inconveniences of, 10–11
  physical properties of, 24–27, *table* (#5) 34
  sizes, *table* (#4) 23, 33, 102–103
    for furnace or boiler, 138
  storage, 4, 31–33 and *illus.*
  supply, 2–4
    purchasing, 27–31
  types, 16–19, *table* (#3) 17
    choice of, 102–103, 115
    ignition temperatures, *table* (#6) 35
    ratings, *table* (#5) 34
  weight and density, 32
  and wood compared, 2, 4, *chart* 5, *chart* 12, *table* (#5) 34
Coal bin, 32–33 and *illus.*
Coal furnace, *see* Furnace
Coal gas, 19
Coal heater, *see* Heater
Coal stove, *see* Stove
Coke, 19, *illus.* 20
  burning procedure, 109
Combustion, 34–36; *see also* Firing stove
Cookstove, 60–61 and *illus.*
  manufacturers (list), 181–83 and *illus.*
Cost:
  of coal, 4, 6, 27–29
  of fuels compared, 4–7 and *tables*
Creosote, 4, 116–117
Dealers:
  of parts for used stoves (list), 62
  of stoves, 62–63, 161–64, 166–73 and *illus.*
Delivery of coal, 27–28

Efficiency:
  of coal stove, 35, 52–54
  of heating system, 10
Energy conservation, 7–10
Energy offices (state), 153–55
Environmental considerations, 11–13

"Fiddle stick," 107
Firebox, 49
  temperature maintenance in, 101
Fire extinguisher, 64
Fireplace:
  coal fire in, 54–55
    burning procedures, 114–15
  flue used to vent stove, 84–89 and *illus.*
  replacing, 1

Fireplace *(cont'd)*
  stove insert, 55 and *illus.*, 56, 85–89
  manufacturers (list), 180–81 and *illus.*
  zero-clearance unit, 98–99
Firing stove, 101–121 and *illus.*
  problems (checklist), 119–21
Flash fire test, 13
Floor protection under stove, 70 and *table* (#7), 71–72 and *illus.*
Franklin stove, 56
Friability of coal, 25
Fuels:
  cost comparison, 4–7 and *tables*
  heat comparison, *chart* 26, *table* (#5) 34
  ignition temperatures, 35 and *table* (#6)
Furnace:
  add-on unit, 128, 129, *illus.* 130
  definition, 123
  design, 125–28 and *illus.*
  draft requirements, 138 and *chart*, 139
  hand firing, 139
  installation, 128–30 and *illus.*
  location, 127
  maintenance, 139–40
  manufacturers (list), 177–79 and *illus.*
  multifuel unit, 134–35
  selection, 124–25
  see also Boiler

Gas Sniffer, 96
Grate, 49, 50 and *illus.*, 51
  for fireplace, 55 and *illus.*
  for heater, 59
  size, 33–34

Hard coal, *see* Anthracite
Hearth, 70–72
Heat circulation, 39, 43, 53, 63 and *illus.*, 64, 95
  from furnace, 127
Heater, 59 and *illus.*
  manufacturers (list), 164–66
Heat extractor, 63 and *illus.*, 64
Heating system:
  coal-burning, 123–41 and *illus.*
  efficiency of, 10
Heat shield, 75 and *illus.*
Heat transfer qualities of metals, 45
Heat value of coal, 26
Hopper for coal burner, 135–36
Humidifier for furnace, 127–28

Ignition temperature, 35 and *table* (#6)
  of coal, 4, 35
Installation:
  of boiler, 132–34
  of add-on unit, 133–34 and *illus.*

Index **187**

Installation *(cont'd)*
  of furnace, 128–30 and *illus.*
  of stove:
    cost, 6–7 and *table (#2)*
    in mobile homes, 97–98 and *illus.*
    safety, 13, 67–100 and *illus.*
Insulation, 7, *map* and *illus.* 8, 9–10
Insurance, 69

Kindling, 31, 103, 104, 111, 112
Kitchen range, 60–61 and *illus.*
  manufacturers (list), 181–83 and *illus.*

Lignite, 17
  burning procedure, 113–14

Mining coal, 20–24
  surface (strip), 21, *illus.* 22, 23
  underground (deep), 20–21 and *illus.*
Mobile home, stove installation in, 97–98 and *illus.*
Moisture content of coal, 25 and *illus.*

National Fire Protection Association (NFPA), 13
  recommendations, 70–100 *passim*

Parlor stove, 56–57 and *illus.*
Peat, 16–17
Permit for stove, 69
Pollution, 11–13
Potbelly stove, 58 and *illus.*
Problems in burning coal (checklist), 119–21

Radiant stove, 41–42
Reserves of coal, 2–3, *map* 18
Roof installation of chimney, 91–93 and *illus.*

Safety:
  of furnace, 125
  of installation, 13, 67–100 and *illus.*
    checklist, 99–100
  of stove, 40–41
  see also Chimney fire; Fire extinguisher; Gas Sniffer; Smoke detector
Seam, 16
Size:
  of coal, *table (#4)* 23, 33, 102–103
    for furnace or boiler, 138
  of stove, 39–41
Smoke, 34, 101, 112
  detector, 64
Soft coal, *see* Bituminous coal
Steel stove, 45
Stoker, 135–38
Storage of coal, 4, 31–33 and *illus.*

Stove:
  air leaks, 53
  air supply, 36, 47–48, 95–97 and *illus.*, 101
  breaking in, 103–104
  circulating type, 42 and *illus.*, 43
  clearance from combustibles, 72–76 and *illus.*, *table (#8)* 73
  dealers, 62–63, 161–64, 166–73 and *illus.*
  door, 46–47
  draft regulators, 47–48
  efficiency, 35, 52, 53 and *chart*
    tests, 54
  firebox, 49
    temperature maintenance in, 101
  firing, 101–21 and *illus.*
    problems (checklist), 119–21
  Franklin type, 56
  grate, 49, 50 and *illus.*, 51
    size, 33–34
    see also Ash: shaking
  installation:
    cost, 6–7 and *table (#2)*
    in mobile homes, 97–98 and *illus.*
    safety, 13, 67–100 and *illus.*
  liner, 46
  location, 6, 38–39 and *illus.*
  maintenance, 53, 117–19; *see also* Ash: removal
  materials used in contruction, 44 and *illus.*, 45
  output comparisons, 40
  permit for, 69
  potbelly type, 58 and *illus.*
  radiant type, 41 and *illus.*, 42
  safety of, 40–41
  selection, 37–66
    checklist, 65
  size, 39–41
  as status symbol, 38
  thermostat, 48 and *illus.*, 49
  upright type, 57–58 and *illus.*
  used, 61–62
    parts for, 62
  window in, 47
Stoveboard, 71–72
Stovepipe, 76–79 and *illus.*
  clearance from combustibles, 77 and *illus.*, *table (#9)* 78
Subbituminous coal, 17
  burning procedure, 113–14
Sulfur, 26–27
Sulfuric acid, 12, 27
Summer maintenance of stove, 118–19
Supply of coal:
  individual, 3–4, 27–31
  national, 2–3, *map* 18

Temperature:
    changes, effect on cast iron, 45
    maintenance in firebox, 101
Thermostat on stove, 48 and *illus.*, 49
Tools for firetending, 63
    "fiddle stick," 107
Trailer, stove installation in, 97–98 and *illus.*
Two-fuel system, 11

Upright stove, 57–58 and *illus.*

Wall:
    installation of chimney in, 93–95 and *illus.*
    protection, 72–76 and *illus.*

Wall *(cont'd)*
    clearances, *table* (#8) 73
    heat shield, 75 and *illus.*, 76
Water heating, 61, 64
Window:
    clearance of stove from, 75
    insulation of, 9 and *illus.*, 10
    in stove, 47
Wood:
    burned in coal stove, 117
    and coal compared, 2, 4, *chart* 5, *chart* 12, *table* (#5) 34
    stove, burning coal in, 43

Zero-clearance fireplace unit, 98–99